Mandy Hoffen and a Conspiracy to Resurrect Life and Social Justice in Science Curriculum With Henrietta Lacks

A Play

A Volume in Landscapes of Education

Series Editors:
William Schubert, *University of Illinois at Chicago*
Ming Fang He, *Georgia Southern University*

Landscapes of Education
William Schubert and Ming Fang He, Series Editors

Mandy Hoffen and a Conspiracy to Resurrect Life and Social Justice in Science Curriculum With Henrietta Lacks

A Play

By

Dana Compton McCullough
Georgia Southern University,
Statesboro, Georgia

Information Age Publishing, Inc.
Charlotte, North Carolina • www.infoagepub.com

Library of Congress Cataloging-in-Publication Data:

CIP data for this book can be found on the Library of Congress website:
http://www.loc.gov/index.html

Paperback: 978-1-64802-488-7
Hardcover: 978-1-64802-489-4
E-Book: 978-1-64802-490-0

Printed in the United States of America.

DEDICATION

I

"Each of us is solitary: each of us dies alone: alright, that's a fate against which we can't struggle—but there is plenty in our condition which is not fate, and against which we are less than human unless we do struggle" (Snow, 1959/2013, p. 7).

I dedicate this work to all those who struggle to create a better world.

II

To Robert, Anthony, Matthew,
Mama, Daddy,
Dory,
Cooper, Daisy, Jessie, Violet, Chipper, Teddy, Tate
And the Great Blue Heron that resides on Lake Cumberland
I appreciate your overwhelming love and support
in this journey. I hope in some way that this work will make
a difference in the world, just as your love and support
have made a difference in my struggle and in my world.

III

Many thanks to those students, coworkers, and friends
who brought sunshine, rainbows, unicorns,
and deviled eggs to this journey.

IV

Wear Red October 4:
Honor Henrietta Lacks and her amazing HeLa cells!

CONTENTS

SERIES FOREWORD

Landscapes of Education

William H. Schubert and Ming Fang He

In this book series, we explore panoramic landscapes of education. We invite a wide array of authors from diverse theoretical traditions and geographical locations around the world to ponder deeply and critically undulating and evolving contours of educational experience. We perceive contours of educational experience as landscapes that cultivate and are cultivated by who we were and how we become who we are as individuals and as humanity (Nussbaum, 1997). We engage with complex hills and rift valleys, rocky roads and serene pathways, war torn terrains and flowering gardens, towering trees and wuthering grasses, jagged cliffs and unyielding rocks, flowing rivers and uneven oceans evolving with flows of life that shape our perspectives, modify our ideas, and forge our actions. Building upon John Dewey's (1916) *democratic conception of education* and William Schubert's (2009) *ideals of love, justice, and education*, we perceive landscapes of education not only as schools but also as *gathering places* (Dewey, 1933) for humans to pursue worthwhile living. We honor the poetics of landscapes of education flourishing with divergence, convergence, diversity, and complexity of experience.

We look for authors who can move in new directions. We open dialogue on educational issues and situations of shared concerns. We create a space for educational workers such as public intellectuals, scholars, artists, and practitioners to engage in inquiries into education drawn from multiple

Mandy Hoffen and a Conspiracy to Resurrect Life and Social Justice in Science Curriculum With Henrietta Lacks: A Play, pp. xi–xvi
Copyright © 2021 by Information Age Publishing

perspectives such as art, music, language, literature, philosophy, history, social sciences, and professional studies. We welcome cross-disciplinary, interdisciplinary, transdisciplinary, and counterdisciplinary work. We look for possibilities that are fresh and poetic, nuanced and novelistic, theoretical and practical, personal and political, imaginative and improvisational.

We expand parameters of educational inquiry substantively and methodologically. Substantively, books in this series explore multifarious landscapes wherever education occurs. Such explorations provocatively portray education in schools, workplaces, nonschool settings, and relationships. Methodologically, we encourage diverse forms of inquiry drawing on a wide array of research traditions, approaches, methods, and techniques such as ethnomethodology, phenomenology, hermeneutics, feminism, rhizomatics, deconstructionism, grounded theory, case studies, survey studies, interviews, participant observation, action research, teacher research, activist feminist inquiry, self study, life history, teacher lore, autobiography, biography, memoir, documentary studies, art-based inquiry, ethnography/critical ethnography, autoethnography, participatory inquiry, narrative inquiry, fiction, cross-cultural and multicultural narrative inquiry, psychoanalysis, queer inquiry, and personal~passionate~participatory inquiry (He et al., 2015).

We feature works that amplify the educational value of mass media such as movies, DVDs, television, the internet, comics, news comedy, cell phones, My Space and Facebook, videos, video games, computers, and the World Wide Web. We hope to explore how we learn through such electronic frontiers in vastly new ways with little tutelage. We hope to encourage creative improvising, problem posing, critical inquiring, and joyful learning illuminated in these new ways of learning though electronic frontiers which are often suppressed and repressed in schooling. We hope to acknowledge the power of human beings to learn without lesson plans, manuals, worksheets, standardized tests, acquisitive rewards, or external standards.

We encourage expansions that move beyond Western orthodoxies to embrace landscapes from the Eastern (Asian), Southern (African and Latin American), and Oceanic (islandic) worlds. We especially want to see renditions move into *third spaces* (Gutiérrez et al., 1995) and *in-between* (He, 2003, 2010) that push boundaries, shift borders, dissolve barriers, and thrive upon contradictions of life. It is our intention that the works featured in this series reveal more of the worldwide landscapes of cultures, ideas, and practices that transgress dominant Western ideologies and their corporate and colonizing legacies. These works have potential in developing transcendent theories of decolonization (e.g., Tuhiwai Smith, 1999/2012), advocating the liberty of indigenous language, cultural rights, and intellectualism (e.g., Grande, 2004), shattering *monocultures of the mind* (Shiva, 1993), overcoming perils of globalization, and inventing a better human condition for all.

We highlight activist and social justice oriented research (e.g., Ayers et al., 2009) and personal~passionate~participatory inquiry (e.g., He & Phillion, 2008) that engage participation of all citizens, encourage respect, innovation, interaction, cohesion, justice, and peace, and promote cultural, linguistic, intellectual, and ecological diversity and complexity. We celebrate postcolonial feminist work (e.g., Minh-Ha, 1989; Mohanty, 2003/2005; Narayan, 1997) that explores migration, slavery, suppression, resistance, representation, difference, race, gender, place and responses to influential discourses of racism, sexism, classism, and colonialism. We also feature ecofeminist inquiry that explores the intersectionality of repatriarchal historical analysis, spirituality, racism, classism, imperialism, heterosexism, ageism, ableism, anthropocentrism, speciesism, and other forms of oppression (Mies & Shiva, 1993).

Books in this series focus on the what, why, how, when, where, and for whom of relationships, interactions, and transactions that transform human beings to different levels of awareness to build communities and public spaces with shared interests and common goals to strive for equitable, just, and invigorating human conditions. We seek explorations of the educational aspects of relationships (e.g., family, friendship), international, transnational, or intercultural understanding (e.g., exile, diaspora, displacement, indigenous knowledge), and circumstances of living (e.g., poverty, racism, alienation, war, colonization, oppression, and globalization). We want to see how languages, literacies, communities, homes, and families shape images of life's *mysteries and events* (Ulich, 1955), such as love, tradition, birth, death, success or failure, hopes of salvation, or immortality. These educational dimensions of life dynamically influence and are influenced by life in and out of schools (Schubert, 2010) and in-between (He, 2003, 2010). Through engaging such pursuits, this book series illuminates how human beings improvise lives (Bateson, 1989) and commitments in diverse, complicated, and often contested landscapes of education.

Unlike more definitively crafted book series that explicate inclusions and exclusions with ease and precision, our invitations continuously expand. The depths and breadths of landscapes where we live surpass everyday gaze and complicate static analysis. We showcase books that bring a sense of wonder and surprise, make the strange familiar and the familiar strange, and evoke what we do not expect. We do not narrow or define the topics of this series. Rather, we open doors to new perspectives, diverse paradigms, and creative possibilities. We invite authors to surprise us with their insightful ideas of what has been, what is, and what might be.

This book is a theoretical inquiry into alternative pedagogies that challenge current standardized practices in the field of science education. In this volume, through Mandy Hoffen, a fictional persona, Dana McCullough, the author, explores how stories of Henrietta Lacks become part of a conspiracy to change science education. Mandy Hoffen, however, never expected to find herself in the middle of a conspiracy. As a science teacher of 20 plus years, she worked diligently to meet the needs of her charges, who are currently ninth- and 10-grade biology students in an age of standardized testing. The author also creates imaginary dialogues which serve as the theoretical framework for each chapter. Each chapter unfolds in a form of a play with imaginary settings and events that bring Henrietta Lacks back from the grave to participate in conversations about science, society, and social justice. The imaginary conversations are based on the author's experiences in graduate courses, direct quotations from philosophers of science, historians of science, science educators, curriculum theorists, and stories of students and their study of Henrietta Lacks in a high school biology classroom. The play describes the journey of a graduate student/high school teacher as she researches the importance of the philosophy of science, history of science, science curriculum and social justice in science education. Through reflections on fictional conversations, stories of Henrietta Lacks are examined and described in multiple settings, beginning in an imaginary academic meeting and ending with student conversations in a classroom. Each setting provides a space for conversations wherein participants explore their personal connections with science, science curriculum, issues of social justice related to science, and Henrietta Lacks.

This book will be of interest to graduate students, scholars, and undergraduates in curriculum studies, educational foundations, and teacher education, and those interested in alternative research methodologies. This is the first book to intentionally address the stories of Henrietta Lacks and their importance in the field of curriculum studies, science studies, and current standardized high school science curriculum.

Dana McCullough, the author, creates an innovative form of curriculum inquiry that connects aesthetics, science education, and qualitative inquiry with curriculum studies and illuminates dynamic interactions among bodily praxis, sensual experience of curriculum, and aesthetics of life. Drawing from diverse theoretical traditions such as philosophy of science, history of science, cultural studies of science, science education, curriculum studies and works of literature written by Emily Dickinson, Herman Melville, Alexander Von Humboldt, Ralph Waldo Emerson, and science journalist Rebecca Skloot, Dana McCullough transgresses methodological and epistemological boundaries to represent visual, acoustic, and other sensual experience of complicated and contested curriculum that predominantly relies on written languages. The author has developed innovative lines of inquiry which are grounded in and move beyond scholarly traditions in curriculum studies (e.g., Au et al., 2016; Gershon, 2011, 2018; He et al., 2015; Ng-a-Fook et al., 2016; Paraskeva, 2011; Pinar et al., 1995/2008). The author calls educational workers to transgress boundaries, politicize hopes, liberate potentials, exercise imaginations, and cultivate radical love to fight against all forms of oppression and enact positive educational and social change that fosters equity, equality, freedom, and social justice for all in unjust and contested times.

We applaud with our hearts and souls that Dana McCullough's innovative inquiries help us to keep asking central educational questions about what is worthwhile, why, and for whom as we continue to live our lives in an unjust world. Dana McCullough invites us to see that acting upon those questions helps transcend inquiry boundaries, transgress orthodoxy and dogma, dive into the complexities and contradictions of life, embody a particular stance in relation to power, freedom, and human possibility, and promote a more balanced and equitable human condition that embodies cultural, linguistic, and ecological diversities and pluralities of individuals, groups, tribes, and societies conducive to the flourishing of creative capacities that invigorate intellectual, emotional, sensual, moral, and spiritual existence for all.

REFERENCES

Au, W., Brown, A. L., & Calderon, D. (2016). *Reclaiming the multicultural roots of the U.S. curriculum: Communities of color and official knowledge in education*. Teachers College Press.

Ayers, W., Quinn, T., & Stovall, D. (Eds.). (2009). *Handbook of social justice in education*. Routledge.

Bateson, M. C. (1989). *Composing a life*. The Atlantic Monthly Press.

Dewey, J. (1916). *Democracy and education*. Macmillan.

Dewey, J. (1933, April 23). Dewey outlines utopian schools. *New York Times*, p. 7. Also in Boydston, J. A. (Ed.), (1989). *The later works (1925–1953) of John Dewey* (Vol. 9, pp. 136–140). Southern Illinois University Press.

Gershon, W. S. (Ed.). (2011). Sensual curriculum: Understanding curriculum of and through the senses [Special Issue]. *Journal of Curriculum Theorizing, 27*(2).

Gershon, W. S. (Ed.). (2018). *Sound curriculum: Sonic studies in educational theory, method, & practice* (Studies in Curriculum Theory Series). Routledge.

Grande, S. (2004). *Red pedagogy: Native American social and political thought*. Rowman & Littlefield.

Gutiérrez, K. D., Rymes, B., & Larson, J. (1995). Script, counterscript, and underlife in the classroom: James Brown versus *Brown v. Board of Education*. *Harvard Educational Review, 65*(3), 445–471.

He, M. F. (2003). *A river forever flowing: Cross-cultural lives and identities in the multi-cultural landscape*. Information Age.

He, M. F. (2010). Exile pedagogy: Teaching in-between. In J. A. Sandlin, B. D. Schultz, & J. Burdick (Eds.), *Handbook of public pedagogy* (pp. 469–482). Routledge.

He, M. F., & Phillion, J. (2008). *Personal~passionate~participatory inquiry into social justice in education*. Information Age.

He, M. F., Schultz, B. D., & Schubert, W. H. (Eds.). (2015). *The SAGE guide to curriculum in education*. SAGE.

Mies, M., & Shiva, S. (1993). *Ecofeminism*. Fernwood.

Minh-Ha, T. T. (1989). *Woman, native, other: Writing postcoloniality and feminism* (Midland Books). Indiana University Press.

Mohanty, C. T. (2005). *Feminism without borders: Decolonizing theory, practicing solidarity*. Duke University Press. (Original work published 2003)

Narayan, U. (1997). *Dislocating cultures: Identities, traditions, and third world feminism*. Routledge.

Ng-a-Fook, N., Ibrahim, A., & Reis, G. (2016). *Provoking curriculum studies: Strong poetry and arts of the possible in education*. Routledge.

Nussbaum, M. (1997). *Cultivating humanity: A classical defense of reform in liberal education*. Harvard University Press.

Paraskeva, J. M. (2011). *Conflicts in curriculum theory: Challenging hegemonic epistemologies*. Palgrave Macmillan.

Pinar, W. F., Reynolds, W. M., Slattery, P., & Taubman, P. M. (2008). *Understanding curriculum: An introduction to the study of historical and contemporary curriculum discourses*. Peter Lang. (Original work published 1995)

Schubert, W. H. (2009). *Love, justice, and education: John Dewey and the Utopians*. Information Age.

Schubert, W. H. (2010). Outside curriculum. In C. Kridel (Ed.), *Encyclopedia of curriculum studies* (pp. 624–628). SAGE.

Shiva, V. (1993). *Monocultures of the mind: Perspectives on biodiversity and biotechnology*. Zed Books.

Tuhiwai Smith, L. (2012). *Decolonizing methodologies: Research and indigenous peoples*. London: Zed Books. (Original work published 1999)

Ulich, R. (1955). Response to Ralph Harper's essay. In N. B. Henry (Ed.), *Modern philosophies of education, Fifty-fourth Yearbook (Part I) of the National Society for the Study of Education* (pp. 254–257). University of Chicago Press.

ACKNOWLEDGMENTS

This work, written as a play, would not have been possible without wonderful people cast in roles of support and guidance. John Weaver challenged me to see science in ways that I had never imagined, introducing me to many individuals and stories along the way that I would never have associated with science. These new associations have informed my teaching and enriched my life. Ming Fang He provided inspirational sunshine and encouragement that not only nourished my brain, but also my soul. James Jupp asked me challenging questions that helped me to view my own work from yet another point of view in this multiperspective endeavor. David Blades' constructive feedback and constant suggestions of resources have furthered this work beyond what I imagined it could be and allowed me to see how this work can be extended in the future. William Schubert made time to read and discuss my work during the early stages, and his suggestions and comments continue to provide encouragement for my work. Mary Catherine Hager read, answered my questions, and provided valuable editorial advice. I would also like to thank my fellow graduate students, friends, coworkers, and students, and random people in grocery store lines and doctor's offices who have listened to my stories about Henrietta Lacks and shared stories they found to be inspirational.

PROLOGUE

Mandy Hoffen never expected to find herself in the middle of a conspiracy. A science teacher of 20 plus years, she worked diligently to meet the needs of her charges, who currently include ninth- and 10th-grade biology students, ninth- and 10th-grade students who have been schooled in an age of standardized testing. Mandy feels that this process of testing her students to death must change. High-stakes testing has overtaken teaching and learning for all participants in education in her district, her state, her country, and her world! Mandy was constantly searching for something that will grab her students' interest and bring life back into her classroom and into the mandated standardized curriculum, a resurrection of sorts. Mandy decides her best move toward changing this phenomenon is to become part of a conspiracy. Don't worry, this hard-working teacher is not joining in on any evil plot that might keep her from earning her teacher retirement pension. Tom Barone (1990) defines conspiracy not in an "obvious and shallow political sense" or with "connotations of evil or treachery as inherent in an act of conspiracy" but as a "profoundly ethical and moral undertaking" (p. 313). Mandy was constantly seeking to find something that would ignite a love for learning in her students. She wanted her students to have a deep understanding of how the science she was teaching them connected to the real world. Mandy wanted each of them to understand that they are stakeholders and participants of science. This was all something she could share with them, but she needed a real-world example. She needed a story. She didn't know it in the beginning, but she needed Henrietta Lacks and the story of Henrietta's amazing HeLa cells.

Mandy Hoffen had never heard of Henrietta Lacks. During a graduate class, a professor asked if she knew about Henrietta Lacks. She had to con-

Mandy Hoffen and a Conspiracy to Resurrect Life and Social Justice in Science Curriculum With Henrietta Lacks: A Play, pp. xix–xxxv

fess that she had never heard her name. Mandy ordered Rebecca Skloot's 2010 book, *The Immortal Life of Henrietta Lacks*, right there in class. The book arrived at her home 2 days later. She started reading and could not put it down. She soon learned that cells taken from Henrietta became the first cells to grow outside the human body, the first human tissue culture. Henrietta was a patient at Johns Hopkins Hospital in the public ward. Hopkins was one of the few places that an African American woman could receive medical care in 1951, when Henrietta was diagnosed with cervical cancer. During surgery, doctors took her cells and sent them to George Guy's lab in the basement of the hospital. Unlike any other sample of cells sent to this lab, Henrietta's cells grew like wildfire, overflowing the test tubes. The lab technician, Mary, labeled these tubes with the letters H-e-L-a—He for Henrietta and La for Lacks. These cells became the first human cells to grow outside the body. The Henrietta Lacks cell line is still in existence today. Henrietta's cells have been to outer space. Henrietta's cells have been used to test and develop the polio vaccine, chemotherapy, cloning, gene mapping, and in vitro fertilization. Her cells have become very important in testing drugs to treat herpes, leukemia, influenza, hemophilia, and Parkinson's disease. Don't you know Henrietta and her family were thrilled about these amazing events! Well, they might have been, had they known. Doctors felt that it was best if Henrietta's family never knew about the cells. Henrietta died in that public ward on October 4, 1951, and the HeLa cells were reported to have come from a woman named Helen Lane. Eventually, the identity of the HeLa cell donor would be resurrected. Stay tuned. This is a story with many plot twists and turns. Mandy was quite taken with the story; some may have even used the word obsessed. She began sharing the story with anyone who would listen, her students, coworkers, family, random people in grocery stores. She felt the story needed to be shared with everyone. She wondered, would we be alive today without Henrietta Lacks and her amazing cells? Mandy set out on a journey to figure out how to share Henrietta's story. She wondered how such a story might impact her teaching, her students. At first it was all about the science, but as she dove deeper into the story of the cells, she realized this story was so much more—a "profoundly ethical and moral undertaking" (Barone, 1990, p. 313).

CONSPIRACY: A FOUNDATION
FOR BUILDING BRIDGES AND LINGERING

Before we begin out journey, let's take a closer look at Mandy's journey into conspiracy. Barone (1990) tells us that a conspiracy is "a conversation about the relationship between present and future worlds" (p. 313). In order to respect this definition of conspiracy, the stories of Mandy and Henrietta will be told in the form of a play, which is for the most part a series of imaginary conversations that take place in some real and some not so real settings. By bringing together the coconspirators, the reader, you, "a historically situated self, learns from the recreated other in the text to see features of a social reality that may have gone previously unnoticed" (p. 314). Join us, you may learn something that you never knew. Mandy, groups of scholars, teachers, and students meet and converse in several imaginary settings. All boundaries of time become blurred. Spaces between old and new worlds are forged.

In order to participate in this conspiracy, all the participants must find a space where these stories can be contemplated. Barone's version of a conspiracy creates gaps, bridges, and third spaces. These bridges occur through the telling of stories, and the witnessing of conversations. I use the word bridge to designate a space where a third thing is created between the participants themselves and the curriculum. Aoki (1996) writes about a type of bridge in which "we are in no hurry to cross over; in fact, such bridges urge us to linger" (p. 5). When we make a conscious decision to linger, we may choose to contemplate the past, present, and future in reference to ourselves, society, or others. This might evolve into a rather complicated story that can be shared. Stories of this nature require a "teller's thoughtfulness, canniness, sensitivity and talent" (Coles, 1997, p. 93). Multiple perspectives can be woven together. When a story is woven successfully, the product that emerges is a kind of truth:

> An enveloping and unforgettable wisdom that strikes the reader as realer than the truth, a truth that penetrates deep within one, that leaps beyond verisimilitude or incisive portrayal, appealing and recognizable characterization, and lands on a terrain where the cognitive, the emotional, the reflective, and the moral live side-by-side. (Coles, 1997, p. 93)

Can this lingering or search for truth be considered research? When stories are part of research one must kept the following in mind: "One can get some 'news,' make some 'observations,' obtain some 'data,' conduct some interviews ... wrap up one's project and leave" on time and on budget or "one can linger and try to learn something other than the answer to one's original inquiry" (Coles, 1997, p. 75).

The word *linger* strikes me as the secret to getting at the heart of the matter. Come and linger a while. "We bring all our sorted histories, hopes, and desires to the project of curriculum theory, hooking onto familiar stories and creating new ones" (Baszile, 2010, p. 483). One might say stories[3] (stories cubed): Their story, your story, and my story—all-intersecting, each of us carrying our own subjectivities and personal histories, each of us being an "other." Coles (1989) says, "It's what we all carry with us on this trip we take, and we owe it to each other to respect our stories and learn from these stories" (p. 30). Our personal stories are part of us. These stories help us make connections as we journey through life. As we travel, we add to our story, making meaning by connecting our stories to new knowledge we acquire as we live our lives. As students are introduced to the story of Henrietta Lacks, they may write and rewrite parts of their own story.

For each participant, writer and reader included, this conspiracy will be a "cautious and wary" endeavor (Barone, 1990, p. 314). This act "ultimately resonates with the interior vision of the text and is persuaded of its usefulness," and can be "borrowed for his own" (p. 314). The text of this work will be woven together in order to create a space for different perspectives concerning science, science education, and possible changes within all that encompasses education to be considered—changes on multiple levels of education involving our character, Mandy Hoffen, high school students, teachers, and society as a whole. The word conspiracy is used to address that this journey will be composed of several groups of individuals all working for the same goal. Each group of individuals brings a different perspective. Barone (1990) says, "There is breathing together, a sharing of ideas and ideals for the purposes of an improved reality. This conspiracy is a plot against inadequate present conditions in favor of an emancipatory social arrangement in the future" (p. 314).

In schools, Mandy Hoffen envisions an improved reality as one that focuses more on individuals, students, teachers, and scholars and less on methods, standardized testing, and marginalization due to test scores. As a teacher, she must find a platform that will provide the space "from which we launch educational projects that aim to change what is into what ought to be" (Barone, 1990, p. 314). I invite you to join Mandy Hoffen and Henrietta Lacks in this conspiracy and become part of the resurrection so desperately needed in education. Be forewarned, a conspiracy is not an easy journey. If you decide to come along, I warn you to be prepared. This text will be full of disruptions ... one minute you might be reading narrative ... the next you will be reading a play. The narrative disruptions are intentional—intentional for the sake of introducing noise into the curriculum. This "noise" is introduced and situated between two loci disturbing each opposing end (Serres, 1982/2007, p. 14). This in-between third place is a "space of transformation" (Aoki, 1996; Rutherford, 1990; Serres, 1982/2007, p. 73). It is in the learning that all things can possibly become new. It will be in this space that we can understand how the history, culture, philosophy of science, science education, and social justice intersect with Henrietta Lacks and participants of science education.

RESURRECTION CREATES NEED FOR A CONSPIRACY

The term *resurrect* in the title of this work was selected from a multitude of words that all mean change: reconceptualize, restructure, rebuild, revive, renew. *Resurrect* is most used to describe bringing back to life, or to rise from the dead. This can mean physical resurrection or the resurrection of ideas. There are multiple examples of resurrection in the work that will follow. Rebecca Skloot (2010) resurrects Henrietta's story from the grave with her book, *The Immortal Life of Henrietta Lacks*. Why did Skloot resurrect this story? As I said before, doctors took a sample of Henrietta's cells without her permission. Doctors at Johns Hopkins University in Baltimore, Maryland, conveniently failed to tell others that the cells came from Henrietta Lacks. Skloot as a biology student was amazed and appalled that no one knew the name Henrietta Lacks. Skloot wrote her book to explain the details of how the Lacks family comes to know that Henrietta's cells were taken without permission and used in medical research. Without Henrietta's cells, many advances in science would be non-existent. In the following pages, Henrietta Lacks herself is resurrected from the grave so she can take part in discussions concerning the intersections of science, philosophy, history, culture, and education. Henrietta's cells have been part of science since 1951: Can Henrietta herself, and her story, also become part of science education?

Henrietta and her story can serve as symbols of the resurrection that needs to happen in present-day schools. Why is there a need for resurrection in schools? Standardized testing has caused many students and teachers to become disillusioned with what goes on in classrooms each day.[1] For example, in my district, classroom instruction is predetermined by a list of standards, posted on the concrete-block wall of my classroom. Not only are the standards predetermined, the method in which instructional delivery takes place is highly monitored and scripted, according to the Teacher Keys Effectiveness System—the current evaluation system in place in the state of Georgia. Not only do students participate in standardized instruction, they are also required to take a standardized end of course assessment for each course for measuring teacher performance. End of course assessments are created by the state. Why all these tests?

The major focus of our most current educational improvement initiative is driven by the Every Student Succeeds Act which replaced the previous "federally controlled No Child Left Behind Act (NCLB), with a more flexible state controlled educational program" (Zinskie & Rea, 2016). ESSA (Every Child Succeeds Act of 2015, Public Law No. 114-95, S.1177, 114th Cong, 2015) requires states to develop and implement accountability systems in order to support student learning. ESSA allows school districts, school staff, and parents to work together and have an "opportunity to replace the one-size fits all remedies of No Child Left Behind with locally selected and designed evidence-based interventions that are creatively adapted to the particular needs of their struggling students and schools" (Zinskie & Rea, 2016, p. 2). With this increased responsibility at the local level, educational decision makers and "schools have a greater responsibility to meet the needs of all students (p. 2)

Under NCLB and ESSA most students were/are enrolled in seven courses. As school districts transition from NCLB to ESSA the amount of standardized testing should change. During NCLB, for each course students take state created end of course tests (EOC) or teacher or district created student learning objective (SLO) tests along with one benchmark per course per nine-week grading period. Benchmark dates are predetermined and fall near the end of each nine weeks grading period. Benchmarks are scheduled based on the curriculum map. These computer-generated tests are composed of 25 or more questions and must be given in the same manner as a standardized test. The length of the test requires the use of an entire class period. Four benchmarks result in the loss of 4 days of instruction.

That added up to over 10 mandated assessments per course, depending on the mix of EOCT's and SLO's distributed throughout the student's 7 period day. This simply means that students may possibly have up to 70 assessment periods in one academic year of school. Most assessment periods involve at least one class period. Assessment results are available for benchmarks and should be revisited with the students: That calculates to another 28 hours just spent on EOCT and SLO testing alone. This calculation does not include teacher-given assessments for the student's coursework.

Transitioning from NCLB to ESSA will take time. In the state of Georgia SLO testing has been removed, however there is still a tremendous amount of testing for all students and teachers. Teachers cannot ignore the importance of mandated testing. This can lead to meaningful activities such as science labs falling by the wayside. Students are arriving in science classrooms "having had little or no firsthand experiences with science" (B. Berry, 2007, p. 5). In elementary school and middle school "they have read about animals and plants, but the scripted curriculum ... does not allow time for hands-on experiments" (p. 5). Testing and practice for testing monopolize all classroom activities. B. Berry (2007) reports the following:

> Students do not know how to pose a question, make a hypothesis, or conduct an experiment. What is worse is that they are not excited about science—they have missed out on the idea that they themselves can investigate

the world around them and make discoveries for themselves, instead of only reading about the discoveries of others. (p. 5)

Instead science students must focus on memorization and becoming accomplished multiple-choice test takers.

Current practices in all subject areas and all grade levels continue to be exemplified by what Freire described as an *educational banking system*. Freire (1970) tells us "the student records, memorizes, and repeats these phrases without perceiving" what the information means (p. 71). Instead of education being an active, enriching, student-centered process, "education thus becomes an act of depositing, in which the students are the depositories, and the teacher is the depositor. Instead of communicating, the teacher issues communiques and makes deposits which the students patiently receive, memorize and repeat" (p. 72). The only opportunity available for students in this situation is for them "to become collectors or catalogers of things they store" (p. 72). Unfortunately, in this scenario, "it is the people themselves who are filed away through the lack of creativity, transformation, and knowledge in this (at best) misguided system" (p. 72). Freire (1970) stresses that "apart from inquiry, apart from the praxis, individuals cannot be truly human" (p. 72). Knowledge is not going to emerge from a system with so many restraints. It will be through "invention and re-invention, through the restless, impatient continuing, hopeful inquiry human beings pursue in the world, with the world, and with each other" that will result in the emergence of knowledge (p. 72).

Heese (2015) provides a detailed account of 24 separate research inquiries that looked at the impact of high stakes testing on curriculum and instruction as related to teacher centered instruction and rote memorization alone. Heese (2015) provides other evidence to tell us what we as educators in a high-stakes testing environment already know: High stakes testing increases teacher stress and anxiety, decreases teacher morale, impacts teacher relationships with other teachers and with students, increases amount of work required in the school day, and diverts funding away from instructional materials and to materials needed for testing and testing preparation. High stakes testing impacts the number of classroom assessments and the formats of how exams are written as well as how exams are weighted in a student's overall grade. High stakes testing depletes time from classroom instruction and activities. Heese (2015) summarizes 27 research articles that deal with loss of instructional time. The evidence is overwhelming that high-stakes testing is not good for education. A one-size-fits-all system cannot meet the needs of our diverse populations in schools. It is no wonder that Grumet (1999a) describes "standardized education as being very short of the rich and abundant experience that education can bring" (p. 24). Curriculum theorist William Pinar (2004) calls this a nightmare:

> The nightmare that is the present—in which educators have little control over the curriculum, the very organizational and intellectual center of schooling—has several markers, prominent among them "accountability," an apparently commonsensical idea that makes teachers, rather than students and their parents, responsible for students' educational accomplishment. (p. 5)

Educators are in dire need of alternative approaches and pedagogy that will resurrect an enthusiasm for learning and bring life and love for learning back to classrooms. Can curriculum studies offer hope for educational changes? Can curriculum studies inform change in the curriculum presented in this teacher's high school classroom? Current conversations in education are being monopolized by standardized tests and standardized curriculum. A recommitment to conversation involving curriculum theory and educational practices needs to take place.

Can including Henrietta Lacks and her story in the classroom breathe life and social justice back into a standardized curriculum, tested-to-death students, and weary teachers? Henrietta has a history of being heard from the grave. These stories are scattered through the text of Rebecca Skloot's 2010 book, *The Immortal Life of Henrietta Lacks*. As two cousins lowered her coffin into her grave and began covering Henrietta with dirt,

> The sky turned black as strap molasses. The rain fell thick and fast. Then came long rumbling thunder, screams from the babies, and a blast of wind so strong it tore the metal roof off the barn below the cemetery and sent it flying through the air ... the wind caused fires that burned tobacco fields. It ripped trees from the ground, blew power lines out for miles, and tore one Lacks cousin's wooden cabin clear out of the ground, threw him from the living room into his garden, then landed on top of him, killing him instantly. Years later, when Henrietta's cousin Peter looked back on that day, he just shook his bald head and laughed: Hennie never was what you'd call a beatin-around-the-bush woman.... We shoulda known she was tryin to tell us somethin with the storm. (Skloot, 2010, p. 92)

What can a reading of Henrietta's story tell us about science and science education? Will curriculum writers of science allow a story such as Henrietta's to enter their conversations and become a component of contemporary conversations about science and science education? Are stories such as Henrietta's deleted in the processes of standardizing science and science education? "If the cultural studies of science seek to become a viable alternative to traditional enlightened science, it will have to begin to invent a pedagogy that offers alternative ways of seeing nature, science and the world" (Weaver, 2001, p. 18). By viewing science through the eyes of Henrietta Lacks, new pedagogy can evolve. The new curriculum should allow students to create space to explore connections between the curriculum and science, the curriculum and society, and the curriculum and the student. In this space there will be stories shared, told, and created.

A METHOD FOR USING STORIES IN THE CLASSROOM

Barone (2000) verifies that in this journey of self-creation through hearing stories and weaving them into our own stories, there will be "uncertainty" (p. xi). Barone (2000) speaks of this journey of curriculum, sans the blueprints" (p. 10) as one that focuses on the two-way interaction between the teacher and student, and their interactions with a curriculum. He says that

> [this] kind of curriculum theory ... will arise from the real qualities of the students' experiences of and their interaction with, the ongoing activities, and from the meanings that facets of the curriculum hold for them ... curriculum from the students' perspectives. (p. 53)

Standardization cannot work here because "[an] individual defines the world from his own perspective" (p. 56). In my classroom, I want to be what Freire (1970) calls a "humanist, revolutionary educator," putting forth efforts that "must coincide with those of students to engage in critical thinking and the quest for mutual humanization" (p. 75). For teachers to achieve these goals, we must work side by side and build meaningful relationships with our students. Palmer (1998) describes good teaching "as an act of hospitality toward the young, and hospitality is always an act that benefits the host even more than the guest" (p. 51). In other words, all those difficulties we face in our classrooms are made worthwhile when we see our students creating their own love for learning.

Barone (2000) tells us that: "All great literature ... lures those who experience it away from the shores of literal truth and out into uncharted waters where meaning is more ambiguous" (Barone, 2000, p. 61). This research will reflect student responses to *The Immortal Life of Henrietta Lacks*. Barone (2000) suggests that, "the reader must imaginatively construct [their] own reality of what [they] read" (p. 61). We must have not only a method for narrative analysis, but also a method to generate information for analysis, and a process for gearing narrative analysis toward science. How can researchers have students and teachers create stories based on a specific science topic? How does the story composition process begin? I am convinced that the story must come first ... and then the careful narrative analysis or interpretations. The issue is that this process may be different for every participant involved in hearing a story. For example, knowing how DNA is sequenced in a laboratory does not mean much unless students know how this DNA sequence impacts them personally and how this knowledge can impact the world.

Barone (2000) offers some advice on how interpretation of text takes place, which can be converted into a "loose" method for sharing stories. First, we teachers must "invite our students into the dangerous vessel which will float them away from the safety of literal truth and the twin seductions of ethical sloth and moral intolerance" (Barone, 2000, p. 61). This invitation will arrive in the form of a well-told story, offered by a very prepared and intriguing storyteller. For instance, the students can read an excerpt and then be asked to connect it to the standards being studied. Next, "we must design activities that entice them into paying careful attention to the social and empirical world around them" ... leaving "gaps for students to fill in, holes which encourage them to actively intervene in the proceedings to assume responsibility" (Barone, 2000, p. 62). It will be the gaps or transformative spaces that provide an opportunity for the students "to think critically about the significance of that which they have experienced, to wonder how it fits into their own maturing outlooks on the world" (Barone, 2000, p. 62), Third, students need an opportunity "to share their tentative thoughts with teachers and each other, to tear down and construct again any conclusions reached and then to act....stay tuned in to the world and ready to change it" (Barone, 2000, pp. 67–68).

When using Barone's approach, students not only get an opportunity to examine the text for specific content related objectives, they will be examining themselves in relation to the story. Pinar (1994) calls autobiography an "archaeology of self" (p. 202). He explains that our life from infancy through childhood "remains hidden from view" but the memories and experiences form "layers of sedimentation—social, private, of various modalities and categories that constitute a self" (Pinar, 1994, p. 202). The experience of autobiography allows each layer to be uncovered one at a time: "Autobiography is a first person and singular version of culture and history as these are embodied in the concretely existing individual in society in historical time" (Pinar, 2004, p. 38). Not only are the layers uncovered, it takes effort to dig through exterior feelings and attitudes for the exploration to occur. Miller (2005) says, "The new ways of knowing can be strange, alien, and frightening" (Miller 2005, p. 76). The process of rethinking all we have ever known may leave us feeling a little ignorant, but we continue to grow and grow—never returning to the ignorance that was present when we began the journey. This process of self-reflection cannot be taken lightly. It allows each of us to explore our life and make all our "fragmented selves" a completely understood self (Miller, 2005, p.

76). In education this work helps us unearth a multitude of feelings and attitudes from a unique perspective.

Once we, students and teachers, are willing to explore new territory or text, or look at each other and the world with a different lens, real change may take place in all dimensions in those spaces of transformation. For this inquiry, *The Immortal Life of Henrietta Lacks* will provide a foundation or framework on which stories will be constructed. This construction of stories will bring illumination to the question: Can Henrietta Lacks become part of a high school science curriculum?

In order to answer this, we will begin our journey with Mandy Hoffen meeting Henrietta Lacks. We will trace her participation in this conspiracy, first at her school, and then at an academic meeting. Following the events of the meeting, we return with Mandy Hoffen to her school.

SETTING THE STAGE FOR A CONSPIRACY

This book consists of this prologue, seven chapters, and an epilogue. Parts of the book are written as narrative, others as a play, and some in the form of poetry. These various formats not only represent a creative approach to writing, they demonstrate ways that science can be presented, understood, and created in an aesthetic format. This work is polytheoretical. I draw upon the theoretical traditions of philosophy of science, history of science, cultural studies of science, science education, curriculum studies and works of literature written by Emily Dickinson, Herman Melville, Alexander Von Humboldt, Ralph Waldo Emerson, and science journalist Rebecca Skloot.

This work weaves together several stories, some fictional, some a telling of actual events. The fictional scripted dialogue shows an evolution of Mandy Hoffen's thinking regarding Henrietta Lacks, science, science education, and social justice. As you read you may notice that not all the characters in this work are fictional. Several scholars agreed to join in on the conversation with actual words extracted from their work, and to have a fictional dialogue created around those quotes. Written permission was acquired. Mandy Hoffen and Delores Pequod are fictitious composite characters. Mandy Hoffen is a teacher who has hope that by using alternative pedagogies in her classroom she can indeed resurrect life in her classroom. I searched for a word meaning hope that could double as a surname. I found the German word *hoffen*. For a conspiracy in education to occur we must have hope. The hope that Hoffen wants to bring to education is very similar to Emily Dickinson's (1993) bird metaphor in her poem Hope (p. 24). Hoffen has hope for education that "perches in [her] soul." She finds this hope at a time when a raging storm of controversy is present in education. One might call this the standardization storm—standardized curriculum, standardized testing and standardized teaching. Rubin and Kazanjian (2011) tell us that

> Standardization and curriculum alignment are the dominant forces in education today. Due in part to the No Child Left Behind Act (NCLB) of 2001, education has become singularly focused on teaching toward the test in order to meet Adequate Yearly Progress (AYP), yet data has shown that using standardized testing does not result in increased student learning or development. (p. 94)

This raging storm involves the detrimental effects that standardization and strict curriculum alignment have, not only on students but on educators as well. Giroux (2009) cried out to our current administration to inform them that they must understand that "the crisis in education is not only an economic problem that requires funds to rebuild old and new schools but also a political and ethical crisis about the very nature of citizenship and democracy" (p. 262). There is "more at risk here than unhappy teachers and over-programed children. There is a battle looming on the horizon as to what type of society we want to live in and what type of citizens will comprise that society" (Rubin & Kazanjian, 2011, p. 103). Mandy Hoffen wants to teach beyond the mandated standardized curriculum. She wants her students to understand their role in a democratic society. She wants them to find meaning in the curriculum that she teaches. She wants them to understand that it is through learning that we can facilitate change in society. She has hope for students and fellow teachers that the standardized curriculum can become infiltrated with meaning that will translate into meaning relevant to life and living. Can her hope provide warmth, even in the "chilliest of lands" of standardization and controversy (Dickinson, 1993, p. 24)?

One of Mandy's adversaries is Delores Pequod. Herman Melville named the ship in *Moby Dick* the *Pequod*, a ship "named for the once-defeated Indian tribe … is the mythic incarnation of America: a country blessed by God and by free enterprise that nonetheless embraces the barbarity it supposedly supplanted" (Philbrick, 2011, p. 27). The *Pequod's* captain, Ahab, "has no qualms about exploiting the whale man under his employ" (p. 29). Delores Pequod has similar characteristics. There is foreshadowing throughout Melville's story that warns us of the gloom and doom to come. Delores Pequod, again a composite character, represents the negative aspects of teaching and education. This character resists change in education. She tries to cast a dark cloud on all who work toward change. Her first name means sorrow. It is such attitudes that might thwart Mandy Hoffen's hope of resurrecting life in science classrooms.

This work is informed by work from several fields of study: philosophy of science, history of science, cultural studies of science, science education, and science theater. Works of playwrights, explorers, transcendentalists, and high school students come together to draw upon and reflect these diverse disciplines. The work has evolved into a web of sorts even though, as you read the story of my journey with Henrietta Lacks, you will identify a loose chronological sequence.

The goal of the conversations in the fictitious settings, taking place without the boundaries of time, will seek to challenge the traditions of science and science education. Henrietta Lacks is a character in the play. She will be the main character and topic of the conversations in the upcoming chapters and scenes. Henrietta is resurrected to challenge scientists and science educators to push existing boundaries in order to reconfigure science and science curriculum in a manner that could help science students prepare for life in the 21st century. Henrietta will listen intently and attempt to become part of the conversation, although not everyone in the conversation will be able to hear her.

WHY A PLAY?

In my classroom, I want to tell stories. I want to make sure I share more than a "single story" that might create "incomplete stereotypes" (Adichie, 2009, p. 1). As you have already noticed from the title, I have chosen to write some portions of this narrative as a script for a play. Why a play, you may ask? Script writing affords me opportunities that the format of a simple interview or conversation lack. The work began as a conversation among philosophers, university professors, teachers, and students. The conversation gained depth when I decided to create fictional settings for the conversations. Before long, the conversation had evolved into a play. Of course, I am not the first to do this. Plato (360 B.C./1976) wrote *Meno* "about 390 B.C. or shortly after" (p. 1). In this conversation, Socrates wants Meno to be aware of "a general definition and to correct the logical mistakes the latter makes, such as failing to understand the difference between a definition and an enumeration of particular examples, or including the term to be defined in the definition" (p. 1). The conversation records a basic lesson. This basic conversation shows up later in "Phaedo" and the "Phaedeus" (Plato, 360 B.C./1976, p. 2).

By creating fictional settings, people from different time periods can meet and converse. In order to write dialogue in my research papers and eventually this book, I began reading books that were written in the form of a conversation. Some of these conversations occurred with professors, other graduate students, and with the authors themselves via email and at academic conferences. Thinking about something we read can sometimes become a conversation we have with ourselves. These conversations are very real, although one person is contributing both parts. I found these conversations very valuable in my work. When we think and talk through something, we are working to understand, and it is through this thinking process that the conversation can inform our thinking as much as an actual conversation. I found that I learned much from real conversations, conversations I had with myself, and conversations that I created fictionally.

HOW TO WRITE A PLAY

Scenes written for Act I were based on actual conversations and actual individuals in my graduate program. Writing these scenes involved a simple recounting of events. Act II-Act VII are composed through academic and imaginative efforts. As I mentioned earlier, I was encouraged to write a "creative" literature review for my dissertation inquiry class. When I approached John Weaver about this possibility, he suggested I read John Casti's book, *The Cambridge Quintet* (Casti, 1998). He also suggested I read Patti Lather and Chris Smithies' book *Troubling the Angels* (1997). I began reading these works and went back to my bookshelf and pulled other works written as conversations. I liked how the conversations flowed together. I wondered if I could pose a question and have several authors respond. Once I started thinking about writing in the form of a conversation, my entire writing process changed. I had always recorded notes in margins. These notes matched up chronologically as I read, page by page, book by book. When I began the process of writing a conversation, I started by posing a question for the participants of a chapter to discuss. Then I would read an author and record quotes that related to the answer to the question. Once I formed the academic layer of the conversation, I began composing and weaving the imaginary layer of the play—the setting, the interruptions, the basic parts of the conversation that told the logistics of the meeting, and made the necessary connections to Henrietta Lacks. The action that occurs surrounding the conversations such as Henrietta rising up from the spilt beaker of HeLa cells and her mysteriously arriving at the conference is based in science fiction. The use of science fiction in this work and its importance in curriculum will be discussed in the postscript of Chapter 3.

Once the work was underway, I emailed the authors of these works and asked them if they had read *The Immortal Life of Henrietta Lacks*. I explained my project and asked them if they had planned to write about Henrietta Lacks. Then I asked if they felt that sharing the story of Henrietta Lacks was important. Once I read their own words, it was much simpler to create their character in the script. I want to emphasize that writing the fill-in dialogue was the most difficult part of writing the play. I have much anxiety about misinterpreting someone's work by writing something that they would not say or even think. Writing dialogue for people that you do not know is extremely difficult; however, writing dialogue for people that you know, mainly professors from my program, is a terrifying endeavor. Participants were sent copies of the work for review. Authors responded with suggestions and changes were made accordingly.

In this theoretical study written as a play, I will explore how a marginalized, nameless, faceless, African American woman, ravaged by science and society, could find her way into the conversations of science scholars, curriculum theorists, educators, and present-day high school students. I will explore how science can seep into the core, cracks, and crevices of present-day education and society. The purpose of this inquiry is to examine the relationships that evolve between science and all its participants. Can a linear standardized curriculum become multidimensional through the addition of the perspectives and stories from *The Immortal Life of Henrietta Lacks*? Can the stories from Skloot's book provide a bridge for science to be connected to the lives of students, their curriculum, their community, and their world? Can Henrietta's story become a presence in science curriculum at the university level, in high school classrooms, other science classrooms, schools, and venues where policymaking takes place?

CHAPTER SUMMARIES

Each chapter will begin with a prescript, or narrative that tells a portion of Henrietta's story. The prescript will be followed by dialog written as a play. The play will detail the journey of Mandy Hoffen and her experiences with the story of Henrietta Lacks. Characters appearing as themselves will be designated with an asterisk. The conversations are built around quotes from each author's work. Authors participating as themselves offered permission for their work to be used in this manner, with an exception of C.P. Snow in Act II, who is deceased. These characters will have an asterisk (*) by their name to designate that they are speaking for themselves. Quotes in the dialogue will not always be introduced. The dialogue is otherwise fictional. Following the action of each Act, there will be a postscript discussion concerning the ideas that developed through the dialogue. These reflections will be written from the author's perspective as a teacher and researcher.

Chapter 1 focuses on Rebecca Skloot, Mandy Hoffen, and others being introduced to Henrietta Lacks. The prescript portion highlights how

Rebecca Skloot came to learn about Henrietta and her cells. The play portion provides a detailed introduction to Mandy Hoffen and her beginning journey with Henrietta Lacks. Act I chronicles Mandy Hoffen's journey from the graduate school classroom to the school where she works. She is contemplating how to share Henrietta's story with her school, her community, even her world. Mandy shares her feelings about the importance of the story of Henrietta Lacks with her administrators, the teachers in her department, her students, fellow graduate students, and finally Henrietta Lacks herself. The work of Denise Taliaferro Baszile, Michele Serres, Donna Haraway, and William Schubert provide the framework for postscript discussion concerning narrative

Chapter 2 begins with a description of Henrietta as a woman receiving medical care in the Johns Hopkins public ward. The prescript explores the "other" and how we as teachers should make connections with the "other." Relationships with the other need to be forged in special in-between spaces. Can teachers forge relationships with students through the creation of in-between spaces in art and science? In Act II the setting of the play shifts to an imaginary academic meeting being held at Johns Hopkins University. The conference, World Alliance of Science and Science Education, is being held in honor of the 70th anniversary of HeLa cells being grown in culture. It is here that Mandy's bookshelf comes to life and she gets an idea that writing a play about science could be a possible way for her to explore how teachers and students connect to each other and the curriculum. Next, Mandy finds herself in a conversation concerning an art exhibit that simultaneously displays Henrietta Lacks and her HeLa cells. As she sees art and science meeting in this display, she decides to investigate playwriting further. The work of Robert Casti, Ming Fang He, Patti Lather, Robert Lake, Rebecca Skloot, Christopher Innes, David Blades, Alan Brody, and others provides the framework for an inquiry into books written as conversation. Meanwhile there has been a robbery in the conference exhibit hall.

Chapter 3 begins with the story of how Henrietta first traveled to Johns Hopkins for medical treatment. Act III launches with the stories of several individuals who have mystical encounters with Henrietta Lacks on their way to the World Alliance of Science and Science Education meeting. During the opening session at the meeting, Mandy hears the story of Henrietta Lacks as told by four writers, Roland Pattillo, Hannah Landecker, Priscilla Wald, and Rebecca Skloot. These authors not only share Henrietta's story, they offer aspects of her story that need to be included in science and science education for the sake of social justice.

Chapter 4 begins with an explanation of why Henrietta Lacks was known only as HeLa and not Henrietta Lacks. As the conference continues in Act IV, Mandy attends a session in which philosophers and historians of science are given the task of determining what science education and science curriculum should look like in society and classrooms today. The works of Sandra Harding, Bruno Latour, Hans Jorg Rheinberger, and Michele Serres provide the framework for this intellectual discussion. There is a disturbance at the meeting; The session is characterized by noise and interruptions.

Chapter 5 opens with a story of how HeLa cells became the very first cells to be grown in culture. Act V highlights a group of curriculum theorists discussing what science in schools should look like. Works of curriculum theorists Peter Appelbaum, David Blades, Nancy Cartwright, Hannah Landecker, Patti Lather, Priscilla Wald, and John Weaver provide the framework for this intellectual conversation. The group is challenged to create a list of requirements for science curriculum and then compare their work to the work of the philosophers in the previous session.

Chapter 6 introduces Henrietta's daughter Deborah. Deborah and Rebecca Skloot make a trip to Johns Hopkins to see Henrietta's cells. In Act VI, the action moves from the meeting at John's Hopkins to Mandy Hoffen's school. Mandy is meeting with a group of curriculum theorists to discuss her ideas of science and science education and how her teaching practice has been informed by curriculum studies. Works by multiple curriculum scholars including, Angela Calabrese Barton, William Doll, Madeleine Grumet, Ming Fang He, William Schubert, Joseph Schwab, and original work by students provide the framework for this intellectual conversation. High school students provide noise, interruptions, and insight that inadvertently drive the conversation about science education, science, curriculum and social justice. The curriculum theorists visit Mandy Hoffen's classroom in order to determine if theory can indeed meet practice. Students provide their responses to the story of Henrietta Lacks.

Chapter 7 shares a narrative about the first "Wear Red to Honor Henrietta Lacks" event at Mandy Hoffen's school. Act VII is a one-act play that picks up where Act VI left off. The story of Henrietta is relayed in a schoolwide assembly. Students and special guests give their account of Henrietta's story and why it is important to science, science education, and society, and why it is important to them personally. Mandy Hoffen encounters Henrietta and seeks to assess the big picture of how Henrietta fits into science, history, and science education. Some information from previous chapters will be repeated here to represent the student's perspective.

NOTE

1. See Brian Heese (2015) for a complete literature review involving the impact of standardized education on teachers, students, and curriculum.

CHAPTER 1

MANDY'S ADVENTURE INTO CONSPIRACY BEGINS

Prescript: Introducing Henrietta Lacks

Rebecca Skloot (2010) "first learned about HeLa cells and the woman behind them in 1988, thirty-seven years after her death, when she was 16 and sitting in a community college biology class" (p. 2). Her instructor, Donald Defler, was teaching the class about the cell reproduction cycle. On an overhead projector he placed a schematic to describe the process of mitosis. Skloot (2010) describes the diagram as "a neon-colored mess of arrows, squares, and circles with words [she] didn't understand like 'MFP Triggering a chain reaction of Protein Activations'" (p. 2). Defler continued discussing "how mitosis—the process of cell division—makes it possible for embryos to grow into babies, and for our bodies to create new cells for healing wounds or replenishing blood we've lost. We learned that by studying cells in culture" (p. 3). He grinned and spun to face the board, where he wrote two words in enormous print: Henrietta Lacks. Skloot explains that she was waiting for him to continue. Instead he erased the board. Class was over. She wanted to know more about the story. Her head filled with questions. "That's it? That is all we get. There has to be more to the story?" (p. 4). She followed Defler to his office. "Where was she from? Did she know how important her cells were? Did she have any children?" (p. 4). Defler responded by telling Skloot that there was no more.

Mandy Hoffen and a Conspiracy to Resurrect Life and Social Justice in Science Curriculum With Henrietta Lacks: A Play, pp. 1–20
1

Rebecca Skloot was so inspired that she went on to write a bestselling book telling Henrietta's story and the story of her family. I will forever be grateful for her work and commitment to uncover Henrietta's story. If you will, join me in examining this story from a teacher's perspective. Defler was teaching his class the part of the science curriculum that involves cell reproduction. I teach this information every year during the first grading period. By taking a moment to mention the name Henrietta Lacks, this teacher was inviting his students to connect to an otherwise very technical curriculum filled with words like "MFP Triggering a Chain Reaction of Protein Activations" (p. 2). At that time there was not much of the story Defler could tell his students. What about Rebecca Skloot's story? Why was Rebecca the student who followed him back to his office? She was a 16-year-old trying to make up her biology credit—she had failed biology her freshman year. Why did Defler decide to embark upon this vital conversation?

ACT I

The first act of this work seeks to share how Mandy Hoffen was introduced to the story of Henrietta Lacks, presented across five different scenes. First Mandy is introduced to the story of Henrietta Lacks in a graduate school classroom. Next, Mandy sets out to share this story at the school where she works. Mandy proceeds to share the story of Henrietta Lacks with her administrator, her department head, and her science department coworkers.

Act I Scene I: Class Meets Henrietta

Characters in order of appearance: Anne, graduate student; John Weaver*, professor of curriculum studies; Mandy Hoffen, teacher/graduate student; Caroline, teacher/graduate student.

The year is 2011. A group of doctoral students is meeting with their professor, John Weaver. The class begins as students are discussing posthumanism with their professor. The conversation then turns to Henrietta Lacks.

Anne: At what point do we consider something posthuman?[1]

John Weaver: "At its core the posthuman condition implies the merging of humans and machines in order to enhance or improve human capabilities" (Weaver, 2010, p. 11). There are multiple ways to consider what makes us posthuman. We have cyborgs and fyborgs. "The cyborg (cybernetic organism) is the more traditional term created to describe any human who is permanently connected to a mechanical devise such as a prosthetic limb, an organ transplant, a pacemaker, or an altered gene sequence" (p. 11). Computers make us posthuman. Ocular implants make us posthuman; the Internet makes us posthuman. Think about a student with a prosthesis. First, only the privileged student will have what they need medically, because only the privileged will have health insurance, and this might lead to a state-of-the-art prosthesis. Someone who is poor will not have insurance or the prosthesis. Kind of like the family of Henrietta Lacks. Her cells revolutionized medicine as we know it, and her family cannot even afford health insurance. Mandy, you know about Henrietta Lacks, right?

Hoffen: (*She began to squirm uncomfortably in her seat. She was used to Weaver's questions concerning science being directed her way. Caroline kicked her foot under the table to bring an answer out of her more quickly.*) No, I have never heard of Henrietta Lacks. What is a fyborg?

Weaver: "A fyborg (functional organism) is a more recent term and describes more effectively humans whose lives are enhanced because of some form of biotechnology" (Weaver, 2010, p. 11). "A fyborg maintains an intimate relationship with technology but unlike the cyborg the mechanical intersection is not permanent. A fyborg is someone who undergoes regular kidney dialysis, has a hearing aid, wears eyeglasses and perhaps soon benefits from some stem cell procedures" (p. 11). Back to Henrietta. Henrietta was the source of HeLa cells. HeLa cells were the first human tissue to be grown outside the body. The book about her is called the *Immortal Life of Henrietta Lacks*. He holds up a book with an orange cover.

Hoffen: Okay (*still red, still worried her response would disappoint her professor; she slid forward in the large rolling desk chair*). I have heard of HeLa cells. I heard about them a long time ago in a cell molecular biology class. (*At least she thought she remembered. Or maybe this was the response that just sounded better than—"Nope … I don't know about HeLa cells or Henrietta Lacks."*)

Weaver: The HeLa cells came from Henrietta Lacks, and scientists have used her cells all over the world. The cells have been used to create vaccines, cures for cancer. At first, George Gey sent them to other scientists for free. Then they were distributed from a lab at the Tuskegee Institute for about "$25.00, for a tiny glass vial of HeLa cells" (Skloot, 2010, p. 193). What is sad is that Henrietta's family cannot afford health insurance. They never received a dime from the distribution of her cells. More recently, John Moore lost, won, and then lost his rights to his own spleen in the California Supreme Court. His cells have become part of another cell line.

Anne: Wait a minute, you mean these people had no control over what science did with their bodies?

Hoffen: Who wrote the story about Henrietta?

Weaver: Rebecca Skloot—S-K-L-O-O-T.

Mandy Hoffen recorded the author's name. Weaver looked at his watch and knew they must move on. Later that day Mandy ordered the book from Amazon.com. It was the end of the semester and time to get to work on the two 20-page papers she needed to write. She also had to pick a topic for her final project for her undergrad prerequisite class. Several days later, the book arrived. I will take a break and read just a little, she thought. The house was quiet, no one home, why not. A few minutes would be a good break from her papers ... she could go back to her pressing assignments feeling refreshed.

That break did not end until Rebecca Skloot's entire work, *The Immortal Life of Henrietta Lacks,* had been eagerly read. At the time, Mandy did not know that it would be through this text that her research inquiry would germinate, take root, and began to grow. By page three of Skloot's book, Mandy was hooked. Skloot describes the cell by saying "the cytoplasm buzzes like a New York City street" (Skloot, 2010, p. 3). Mandy began to make a list of all the science lessons her students could learn from this book about Henrietta. She continued to read.

In the opening chapter, Skloot tells about Henrietta jumping out of her car and entering Johns Hopkins Hospital. "She scurried into the hospital, past the 'colored' bathroom" (Skloot, 2010, p. 14). Soon Mandy was reading that "[in the] era of Jim Crow—when Black people showed up at White-only hospitals, the staff was likely to send them away, even if it meant they might die in the parking lot" (p. 15). On this day, Henrietta would be seeing Richard Wesley TeLinde, one of the top cervical cancer experts in the country. TeLinde "often used patients from the public wards for research, usually without their knowledge" (p. 29). Many scientists at the time believed that "since their patients were treated for free in the public wards, it was fair to use them as research subjects as a form of payment" (p. 29). Mandy was starting to see that science justified the use of the innocent for its advancement. She could not stop reading. Suddenly the basics of cell biology, memorizing the function of the mitochondria, learning what happens in the nucleus or on the ribosome, did not seem so important. What did seem important was that all the advances made in cell biology were made because a doctor took Henrietta's cells without her permission. These HeLa cells became the first human cell line created for research. More importantly, these HeLa cells came from a real person, with a real family.

Mandy was continually thinking about Henrietta while beginning her journey into the field of curriculum studies, and she knew the two would intersect in many ways. Mandy had always used storytelling in her teaching. This passion began to integrate literature into the required science curriculum. As she began to share excerpts of this story with her students, the basic lessons in science took a back seat. Issues of social justice occupied the forefront of class discussions. Students made connections with Henrietta as a person, Henrietta as a Black woman living in a time of racism and segregation, and Henrietta's identity in relation to science. As a teacher, the standards on the concrete wall were fading ... forced integration of literature for the sake of Common Core became less and less important with each class discussion. Students were taking a stand for social justice as they discussed how Henrietta should have been treated. It was obvious they felt the need to stand up for human beings such as Henrietta Lacks. Mandy wanted to do more than just share Henrietta with her own students. Mandy wanted all the biology teachers to do the same.... She wanted to share the story with the whole school. As time went on, Mandy realized that her reasons for wanting to share were quite complicated. She told her professor that she would love to take her project districtwide ... statewide.... The journey would not be as simple as her love and enthusiasm for Henrietta and her students.

Act I Scene II: Henrietta Lacks Goes to High School

Characters in order of appearance: Mandy Hoffen, graduate student/classroom teacher, and Cathy Osmond, assistant principal

In preparation for planning a schoolwide Wear Red to Honor Henrietta event to introduce the entire school to Henrietta Lacks, Mandy visits the administrator in charge of special events. She taps on the open door.

Hoffen: Mrs. Osmond, I need to talk to you about Henrietta Lacks.

Cathy Osmond: Nope, that is Mr. Hardy's part of the alphabet. He has the L's.

Hoffen: You don't understand. Henrietta is not a student. She is a woman whose cells are very important to science. Have you heard about the book—*The Immortal Life of Henrietta Lacks?*

Osmond: Oh, I am sorry. No, I have not heard about the book. (*Mrs. Osmond was still a little startled that Henrietta was not a student having a discipline issue.*) That sounds very interesting. What exactly do you want to do?

Hoffen: I plan to keep it simple. I want to send some information out to all the teachers for them to share with their students. I want all the students, teachers, and faculty to wear red. The biology teachers are going to use the book, *The Immortal Life of Henrietta Lacks*, to teach a lesson on that day. I was also thinking we would play some swing music at lunch. Henrietta and her cousins would empty all the furniture out of her living room on Saturday night. They would crank up tunes by Benny Goodman and Duke Ellington and dance all night long. Maybe a jazz band, or just some music streaming.

Osmond: Just keep me in the loop. That all sounds fine.

Act I Scene III: Science Department and Henrietta

Characters: Mandy Hoffen, graduate student/teacher; Department Chair, Delores Pequod and Penny Middleton, science teachers

Later that day, Hoffen meets with her science department. She shows the group the copy of the book, *The Immortal Life of Henrietta Lacks*.

Hoffen: I am so excited. Mrs. Osmond has cleared us to have a special day to honor Henrietta Lacks. I read this book this summer (*she holds up a copy*). Henrietta's cells were the first to be grown outside the human body. There is so much information in the book that goes along with our biology standards. Random House has sent me copies of the book, if anyone would like to borrow one. I have posters from Random House for everyone!

Department Chair: Let us know what you want us to do ... but keep in mind ... we don't have a lot of time to spare—lots of standards to cover. I will say though, this is a great way to get those Common Core objectives into your lesson plans.

Delores Pequod: We barely have enough time to teach the curriculum now; how are we supposed to drop a whole day to teach something that is not a specific standard? I might have a benchmark[2] that day.

Hoffen: (*Feeling very defeated upon hearing the negative but true comment.*) I am not asking for a whole period, just a few minutes to tell the students about Henrietta and her HeLa cells. Everything you need will be in a PowerPoint. I know we don't have a lot of time, but I think it is important to share this story. There are so many lessons about ethics in science, and the cells themselves. (*She was obviously rattled. She could feel the other teachers' piercing stares behind her back as she addressed the department head.*) We are all required to document how we are implementing the Common Core standards in our lesson plans. I will send you lesson plans for the event, including the specific standards covered, so you can just pop them into your lesson plans for that day. It is also a great opportunity to connect science to the world. The students might find the science more interesting if they see real-world connections.

Chair: Sounds good. We all must document the standards—this sounds like an easy opportunity.

Penny Middleton: I think it would be great for the students to debate whether the doctors had the right to take her cells without permission. The students could write a short paper supporting one side or the other.

Hoffen: That is a great idea. I also think it would be helpful for the students to see how information about cells in Skloot's book is different from how the information is presented in their textbook.

Pequod: Well this is all well and good for you Biology folks, but I don't really see how it can relate to physics.

Hoffen: The book has many examples. Mrs. Pequod, your students could study the mechanics behind Gey's drum roller. It was crucial for the cells in culture to be in constant motion. Interestingly, he never applied for a patent for the machine. He just passed the idea on to others … all for the sake of science. I think addressing the idea of science and ethics with all our students would be time well spent. There is also the history involved … not only the techniques developed for science, but the fact that the face of medicine changed forever because of this woman's cells. I have a list of science-related topics matched up to passages from the book. I will send those out to everyone.

Like all meetings, most everyone in the room just wanted the meeting to be completed so they could go back to their classrooms and get ready for the next day. The meeting ended. Mandy was very excited, but also very anxious. She went forward with careful preparations. She wanted to give the teachers a simple plan that they could simply carry out. She knew some would participate and some would not.

Act I Scene IV—Dinner Table Discussion

Characters in order of appearance: Anna, graduate student/teacher; William Schubert*, curriculum theorist; Ming Fang He*, curriculum theorist; Cabrala, graduate student/teacher; Mimi, graduate student/teacher; Mandy Hoffen, graduate student/teacher; and Michael, graduate student/teacher

Mandy and her fellow graduate students have gathered at the home of Ming Fang He and William Schubert for lunch following a morning class. It was the last class meeting of the semester and students were enjoying small talk and their time together. Everyone was wearing their slippers—compliments of their host.

> **Hoffen:** I have a question for everyone. You all are always very supportive of my work. I am so grateful to have such a great group. (*She needed to ask this question. Was this the right time? She took a deep breath.*) I need everyone to be very honest with me: When you hear me, as a White woman, tell the story of Henrietta Lacks, do I portray any part of the story in any way that could be perceived as racist? Will my African American students think I am racist when I tell this story? Am I exploiting Henrietta in any way by telling her story?

> **Anna**: "I find the difference in race secondary to the importance of the story. The similarity of gender binds you in challenging the male dominated field of research and medicine. Your position as a teacher to multiple races and genders positions you in a place to widen their worldview as well as emphasize the continuing work necessary in the field of social justice. You are not looking to assume her voice but highlight her struggle and bring it to the forefront for more readers and possible leaders of the community" (A. N. Waddell, personal communication, January 25, 2015).

A voice from the kitchen interjects a question into the conversation.

He: Could a White man not tell this story?

Anna: Good question, Dr. He.

Hoffen: I think anyone can tell this story. Part of the class activity involves every student telling part of Henrietta's story. This will include the White males. I must be prepared to guide them as they tell this story of a Black woman. The activity provides all students an opportunity to examine gender roles, the roles of women, and their own role in society

Schubert: "I would say that you should honestly characterize who you are and why you hold Henrietta Lacks in such high regard, even noting too the possible criticisms and, despite them, why you are moved to make her the centerpiece of your work. You have a highly defensible implicit rationale for doing so and you need to make it explicit. That's my take" (Schubert, 2015).

Hoffen: Thanks, Dr. Schubert.

Michael: "I suggest you simply tell the story and make the focus about Lacks. Write it in a way that if your name wasn't attached as the author, people would perceive the writer as being Henrietta Lacks herself" (M. Williams, personal communication, April 7, 2015).

Dr. He emerges from the kitchen to sit down with the group.

He: As researchers we are "immersed in lives and take on the concerns of people who are marginalized and disenfranchised, and act upon those concerns" (He & Phillion, 2008, p. 2). Just remember that no matter how much time you spend studying them, or how much you know about them, you "cannot become their voice" (M. He, personal communication, July 29, 2014). I think your students will see that your goal is to "advocate for disenfranchised, underrepresented, and invisible groups and individuals … to foster social justice for educational and social change" (He & Phillion, 2008, p. 3).

Cabrala: "I think that you can present it in a way that is logical and rational. I think the telling wouldn't be exploitation if you give a reflexive statement upfront about your position as a White female. If you are candid about this, then I think you will appear transparent and authentic to your audience" (C. Awala, personal communication, April 4, 2015).

Hoffen: Thanks, Cabrala. I want my students to know I am sincere. A doctoral candidate at Emory wrote a letter opposing Rebecca Skloot's portrayal of Henrietta's story. Her name is Rebecca Kumar. She says that Skloot begins the story with "racism on the first pages of the book" (Kumar, 2012, p. 3). Her main objection was that Skloot says in the opening that *The Immortal Life of Henrietta Lacks* is a work of nonfiction. I read it for myself, but honestly did not question. Skloot spent 10 years researching. Skloot does this on the first page, first sentence of the book—let me read it to you. "This is a work of nonfiction. No names have been changed, no characters invented, no events fabricated. While writing the book [she] conducted more than 1,000 hours of interviews" (Skloot, 2010, p. xiii).

The woman objecting to Skloot's telling does not disagree that Skloot was very diligent in writing the book. Kumar says that it is impossible for Skloot to write truthfully about events that she herself did not witness, "the style and content of the story is ultimately the result of her choices … Skloot's descriptions are loaded with political implications and consequences … Skloot has control over the voices" (Kumar, 2012, pp. 3–4). I guess when any of us write, we are taking control of the story we are telling. The passage she objects to from the book describes Henrietta finding her tumor while in the bathtub. How can Skloot know what happened? Why not just say up front that "it could have happened like this" (p. 4)? Kumar also feels that Skloot's nonfiction telling of the story in the voices of Henrietta's family "reads like a Looney Tunes character" (p. 4). Honestly, at first, I chose to brush over the article and ignore Kumar's concerns. The more I thought about it, the more I realized that there will be many opinions about *The Immortal Life of Henrietta Lacks*. Just because it is a best seller, and I think it should be used to teach science and social justice, does not mean that it was written for that purpose alone. Kumar (2012) says that Skloot goes "out of her way to make black life seem strange, funny, and sometimes with her depictions of religion, misguided and uninformed. And this makes [Skloot] the voice of normalcy—authoritative and godlike" (p. 5).

Anna: Interesting; I can see how you cannot ignore this.

Hoffen: Another issue that comes to mind involves victory stories. After my first presentation at the curriculum studies summer collaborative session, Jim Jupp mentioned that when working with students in multicultural settings we must make sure we are not just trying to write a "victory story" (Jupp, 2004 p. 36). I had used the word victory in my presentation; sweet victory … was the expression that I used. It embarrasses me now. I said I felt as if using *The Immortal Life of Henrietta Lacks* helped me connect with my African American students in class. For the first time a group of young men seemed to be interested in biology. Each day I found myself redirecting this group of boys. Jupp sent me an article that discussed how when we are teaching in "multicultural settings, we as White educators can get hung up on our personal victory of making opportunities available to students that are of a different background" (p. 36). Through careful self-reflection, "members of the privileged majority need to reflect and be aware of their intentions and make sure that they are not just going in to save the poor, the needy, and the minority" (p. 36).

Michael: Students who have experienced oppression will see right through this.

Anna: Have you had anything like this come up in your classes? Students are honest. I am sure if they felt your intentions were racist someone would have spoken up and said—how do you know about being a Black woman with cancer.

Hoffen: Students are very honest. My students are very receptive to the story. I am very self-conscious about this ... I feel like the White teacher ... maybe because I have been called this—not in reference to Henrietta ... but when I had to discipline a student who disagreed with her consequences. Do you guys remember reading Robert Coles in Chapman's documentary class?

Anna: I do. I love the story of Ruby Bridges.

Hoffen: Bob Moses, the leader of the Robert Coles research group, writes a very understandable description of White privilege. I wrote it down in the back of my *Immortal Life* book:

> Don't you see, that's been our story—the Black story—everyone calls us something! It is so hard for any single one of us to be seen by you folks [White people], even the kindest of you, even our friends [among you] as a person, nothing more. That is where we are; that is where we are coming from; that is our 'place' in all this! You can be here, and you can be there. You can go set up your tent wherever you think it'll do you good! That is great for you. That's what it means to be White and have a good education. You can look at things with a microscope or a telescope, and from way up in the mountains and down near the seashore, when it's sunny and when it is raining cats and dogs, and then later, when you write or you publish your photographs—you're not a White writer or a White photographer. You're free of the biggest label of them all, the one that defines us every single minute of our lives. (Coles, 1997, p. 40)

This conversation took place during the heart of segregation. I think having these difficult conversations concerning race is necessary. Will we ever get to a point in the South where no one must be sensitive when discussing issues of racism? Will I ever just be the teacher that cares about her students? Can inviting Henrietta into my classroom make a difference in how students view racial issues in school and society?

Hoffen felt like she had dominated the conversation long enough. The conversation picked up concerning other issues with other students' inquiries. To herself she thought: I wish I could talk to Henrietta. She had a story so different from anything I have experienced. How many of us have had our cells sent up to outer space? Henrietta and her parents, who were sharecroppers in Virginia, farmed the same land that their ancestors had worked as slaves. Armed with only a 6th grade education, Henrietta—a wife and mother of five children—entered the world of white coats at Johns Hopkins Hospital, complete with a "colored" bathroom and water fountain and segregated wards, and proceeded to change the course of medical history (Skloot, 2010, p. 13).

Act I Scene V—A Conversation With Henrietta

Characters in order of appearance: Mandy Hoffen, graduate student/teacher and Henrietta Lacks

In the bottom left corner of a dark stage, a teacher sits in a classroom. On the white board in front of the classroom is a painting of an African American woman. The teacher, Mandy Hoffen, is speaking to the painting. Her classroom is empty.

Hoffen: (*Sighing deeply and shaking her head as her seventh-period students rushed out the door.*) Wow, Henrietta, what a day, I feel like we need to have a conversation. I hope you are not offended that I am so taken by your story. Your journey into "the land of white coats" at Johns Hopkins Hospital as a wife, mother, and cancer victim in a time of segregation has opened my eyes to events in history and science that were lacking in my science education (Morris, 2008, p. 11). I realize that so many just know you as HeLa, not Henrietta. I have studied your story so much; I feel like I know you and your family.

Your family believes "your spirit [has] lived on in [your] cells, controlling the life of anyone who crossed its path" (Skloot, 2010, p. 7). Some may feel that this is scientifically impossible, but who am I to question the beliefs of your family and those closest to you? If this is true, your spirit has blessed me. I see great things happen when I share your story with my biology students. Some might think I am crazy ... believing in your spirit. Rebecca Skloot had some doubts about this at first. I like how your daughter, Deborah, set Skloot straight: "How else do you explain why your science teacher knew her real name when everyone else called her Helen Lane" (p. 7)? I cannot explain how I just happen to be in a doctoral program with another teacher that knew your name. Hearing your story changed my life, my teaching, my understanding of science, forever. Is your spirit—along with your cells—immortal? Others hearing your story, sharing your story, writing your story, allow your story to become as immortal as your cells.

I hope you understand that it is my desire to make your story come to life, resurrected from the grave, in the hearts and minds of my biology students. As they live and share your story, maybe you can become immortal like your cells. Your story—complete with issues of science, history, ethics, and race—is full of opportunities for students to situate themselves in another place and time. These young people can make the world know that it is the people like you, Henrietta, who cannot be disregarded in science, or in any other aspect of life and living.

I always ask students how they think your family felt after learning that your cells were still living in laboratories all over the world, 25 years after your death. How could your family deal with your physical death as they were engaged in a "lifelong struggle to make peace with the existence of those cells, and the science that made them possible" (Skloot, 2010, p. 7)? I am constantly reminded of your family's struggle with doctors only calling when they wanted something, strangers calling wanting to ask lots of questions. Was your son Lawrence right when he said that you would want them to see "that there is a positive side to everything, even when things were 100% awful" (Lacks et al., 2013, Chapter One, paragraph 6)? Your cells have helped "millions, if not billions of people all over the world" (Chapter One, paragraph 5). Lawrence imagines that you would be "extremely happy about helping others" (Chapter One, paragraph 5). Your cells became a miracle for humanity—all humans with no regard for race or class—yet no one knows your name.

Lawrence, in speaking of his reason for telling your story himself, says it best: "I want everyone who knows about my mother's cells to know her as a person too. I want everyone to get a sense of her sweetness and humor and spirit despite all the hardships life dealt her" (Lacks et al., 2013, Chapter One, paragraph 9). I agree with Lawrence, but do I have the right to tell others your story? Are you okay with my reasons for wanting to tell your story? I believe that your story can resurrect life in my science classroom. Is this exploitation? So many of my students feel that the science they are required to learn is disconnected from the practice of science. I want the students to feel connected to the science they must learn. I want my students to know that they are part of science. Just like you. Things can get so confused and so misunderstood in our mixed-up, profit-seeking world.

My doctoral research and my work in and out of my classroom is a search for understanding. In this search I will be relying on the words of others. I want to assure you that I will not put blind faith in their words. I want to say up front that it will be a work of fiction. I don't think human composition can be anything but fiction. I seek truth … but truth is complicated. I have learned in my graduate studies that individual narratives, our stories, support the truth we portray. This makes truth a composition. We as individuals are composite characters made of truths and untruths. We choose the truths we want to be a part of ourselves and then we portray those truths in our daily life. The person we see and believe ourselves to be may be very different from the truth that our everyday audience, our family and coworkers, may see. Somewhere between the fiction and truth, subjectivity and objectivity, we will find what we consider our personal reality. In my writing, I have come to accept that everything I represent may be limited to the desired sense of truth I wish to portray. This sense of truth may not be considered an objectively truthful representation because of the subjectivities I bring to each keystroke, each word composed. This truth may be a contradiction to how others perceive the story I write.

With this said, I realize that my accounts of your story as well as the accounts and actions of other characters in my story can only be those of fiction. I will rely heavily on the accounts of your life offered by your family: Deborah, Sonny, Zakariyya, Lawrence, and Bobbette. I will complete a careful study of the book written by Rebecca Skloot, *The Immortal Life of Henrietta Lacks*. I will consider the praises and the criticisms of this text, in order to tell a version of the truth. I specifically say version, because truth can be relative.

My biggest fear in embarking on this journey is that someone may consider me a racist. My desire is not to "perpetuate" or "endorse" the blatant racism faced by your family, "faced by [you] and a countless number of poor women whose bodies were and continue to be exploited in the pursuit of 'knowledge'" (Kumar, 2012, p. 5). As a White woman, I am aware that my life came with a multitude of prepackaged privileges. I was not aware of my privileged life until I returned to school to begin working on a doctorate ... more evidence that I am indeed privileged. I know that crossing boundaries, pushing against preestablished borders of society, and attempting to understand a life lived outside of my race in a distant place and time is a complicated proposition. I hope you will know that you will be the foundation for the resurrection I hope will take place in my classroom and in my teaching, a resurrection in education that I hope will lead my students to an immortal love of learning. I hope the students will understand their role in what becomes immortal in our society. I dream of classrooms, communities, and all of society that are full of social justice for all participants. Don't you think education should cover social grounds just as well, Henrietta? I think my students can see how things in our world need to change because they learn about you and your struggle as a woman living in an era of segregation. Maybe these lessons can become as immortal as those infamous cells of yours!

Hoffen walks to the door, flips off the lights, and leaves the stage. The picture of Henrietta is illuminated, and the audience hears a voice.

Lacks: My name is Henrietta Lacks. The world only knew me as HeLa. Several have made attempts to hide my story. Several have attempted to tell my story. The fact remains that all who hear my story will write their version of it, composing, weaving, and navigating in a sea of complicated issues. I dare you to become part of this conspiracy. I dare you to become entangled in my world, a world of medicine, science, and racism. A world that desperately needs change. Who knows, the changes might begin right here in this classroom.

POSTSCRIPT: MEETING HENRIETTA

This intellectual dialogue over the course of Act I serves as a painter's palate or a chalkboard upon which to compose ideas. The dialogue is multifunctional. It tells a story of a teacher/graduate student/researcher who wishes to share the story of Henrietta Lacks in science class. We see

Mandy starting her graduate studies and working in her school. The curriculum theory she is exposed to in her coursework begins to inform her teaching practice immediately. These conversations bring to light issues that teachers face when wanting to make changes in a standardized curriculum. With the current emphasis on testing and evaluation systems, the priorities are all about standardized testing. Anyone not involved in schools can see how the process of curriculum works in school. Teachers must have the support of their administrators and their department in order to do anything new. This intellectual conversation provides insight to how a doctoral graduate student begins to develop a theoretical research inquiry. Coursework and conversations enable researchers to find their way to their research inquiry. Only in a work of fiction could Henrietta be part of a conversation concerning issues of science, science education, and social justice in schools.

When I examine my personal journey in the research process, I see that sharing Henrietta's story with my students came first. During discussions in my graduate courses and in writing assignments for these courses I was working to find a way to explain how the story of Henrietta Lacks was informing my teaching. I began asking: How could the story of Henrietta and my stories relating my experience teaching students about Henrietta become a research inquiry that could possibly inform curriculum studies? I said that I was grateful to Rebecca Skloot for writing Henrietta's story. I think I also must be thankful to her teacher for introducing his entire class to Henrietta Lacks. The book that resulted, *The Immortal Life of Henrietta Lacks*, has paved the way for the world to hear Henrietta's story. This work of literature provides a pathway for teachers to introduce Henrietta's story to their classes. I know that I cannot just tell my students a story. I must be able to tell a meaningful story and then communicate the results of this telling in this book. How will this telling be done?

Intellectual dialogue and debate can begin with simple conversations about stories. Conversations have a huge role in addressing issues in learning. Can the simple act of allowing teachers and students to give voice to their concerns result in conversations and possibly lead to learning? Wang and Yu (2006) offers the following concerning conversations:

> As we converse, talk, and learn from each other, we come to realize that perhaps there is no definite beginning or end. Wherever one starts, as long as the interactive dynamics between person and structure and between self and other can be kept instead of abandoned, one is open to the creative potential of an intertwining, evolving, and transforming process engaging both self and culture. This process of teaching, conversing and writing does not offer any resolution or formula ... there is no discourse or practice inherently liberatory or empowering, but our pedagogical desires, discourses,

and practices are complicated along the way to reach new possibilities ... going beyond any promise to open up alternative paths. (p. 35)

How does one go about creating opportunities for conversation in the science classroom? How can conversations be used to enhance a curriculum? Remember that curriculum studies is all about reading, writing, and talking.

As I began exploring conversations, I found myself seeking opportunities to write them. Fortunately, my professors were open to alternative styles of representation. I knew that I wanted my work to reach a broad audience. I also knew that I wanted the language to be accessible. Michele Serres (1990/1996) tells us that "technical vocabulary seems immoral: It prevents the majority from participating in the conversation, it eliminates rather than welcomes, and further, it lies in order to express a more complex way things that are often simple" (p 25).

I wondered if my writing about educational theory and my own classroom practices would appear clearer when presented in the form of a story or a conversation. What if science lessons were written as stories? It's worth noting that my writing background up to this point has been truly technical. My master's thesis consisted of five traditional sections—research, hypothesis, method and materials, results, and conclusion—resembling the basic scientific method, 100 graphs, and 100 data tables. Writing conversation was very different. My first attempt at writing a conversation was a discussion between two teachers. My next attempt included a fictional setting. What started out as a simple dialogue became multidimensional because in that format, time and space components can be manipulated. I felt that adding fictional elements would expand the boundaries within which I was working. I would read an author and wonder what this author would say to another ... maybe from works written 100 years apart. Some authors were not in the same field. In a formal style of writing, facts and opinions can be regurgitated. I did not feel that a technical style of writing would allow me to tell my story.

I wanted to take the authors that I was studying, put them in the same place at the same time, and get them to address current issues with science and science education. I could demonstrate how over the course of reading the works of many authors, the reader assimilates the information and uses the information to work through questions or problems. As time continued, the play that I had not intended to write was being written and performed in my head. The play is simply a story with multiple dimensions. Can stories and storytelling be part of teaching, learning, and science curriculum?

NARRATIVE, STORYTELLING, AND SCIENCE

Science is a very complex subject to study. Donna Haraway (Haraway, 1989) tells us that certain aspects of science itself may be considered a kind of storytelling practice: "a governed, constrained, historically changing craft of narrating the history of nature" (p. 4). Within the social constructs of school, teaching, and learning there are many opportunities for stories. I see stories as a more comfortable or informal way to share and converse with others. We share stories everyday through conversations. Can lives be changed with storytelling? "Everything and everybody has a story" (Baszile, 2008, p. 253). Baszile gives some excellent points in learning how to use storytelling in the classroom. Stories can be used as "a primary form of communication," to help "define problems," and to "identify relationships among problems," or they can be employed to help us "laugh at our problems and heal our hurts" and "help us to discover and understand one another" (p. 253). I think these roles of storytelling can be extended to help us to discover and understand science curriculum.

For storytelling pedagogy to be successful, students and teachers must get beyond themselves. Both groups must step back and take a fresh look—to treat oneself as a stranger or an "other." By looking at ourselves as the other, we see the stranger within us. When examining the stranger, we can relate more to the other element in society. This examination "implicates the other in me," such that "my own foreignness to myself is, paradoxically, the source of my ethical connection with others" (Butler, 2005, p. 84). From a curriculum theory perspective, we apply this examination of the other to storytelling: "We bring all our sorted histories, hopes, and desires to the project of curriculum theory, hooking onto familiar stories and creating new ones" (Baszile, 2010, p. 483).

Creating new stories can only come about in classrooms where teachers are willing to address all issues, even those found to be uncomfortable, and allow students to be active participants in the dialogue. For instance, teachers cannot begin to teach about issues of race until they are familiar with their own personal history, and their own personal attitudes concerning race. Good teaching will result when a teacher incorporates results of careful self-reflection into the required standardized curriculum taught in the classroom each day. Although standardized curriculum does not lend itself to change, good teaching will push the boundaries of standardized curriculum in order to bring about change. In Kieran Egan's (1986) book, *The Great Stories of the World Curriculum*, Egan employs "teachers as storytellers" (p. 109). Using great stories in all facets of the curriculum allows students of any age to build a springboard upon which they can begin a process of their own inquiry in any subject. Egan's stories of the world must be based on "science, technology, language, history, life on earth, the stars and planets and so on" (p. 108). Science stories would need to include "what we know about the stories of life on earth, of our place in the universe and of the human ingenuity which discovered the material for these stories" (p. 108). Unfortunately, in our society "science and math are taught as inhuman structures of knowledge ... taking pride in their logical and inhuman precision" (p. 30). Looking at curriculum through a postmodern lens,[3] as well as incorporating stories, may provide pathways to "rehumanize" science curriculum (p. 30). Characteristics of this postmodern curriculum should call into question "rigid dichotomies created by modernity between objective reality and subjective experience, fact and imagination, secular and sacred, public and private" (Weaver, 2001, p. 8). Science pedagogy created within a postmodern framework will "value personal knowledge as much as empirical knowledge and neither places sources of knowledge into an arbitrary hierarchy nor accepts the universal certainty of empirical knowledge" (p. 11).

Relating to the familiar is a comfortable act for students. The challenge in using storytelling in teaching is to teach students to listen and learn from stories in which they feel no connection at all. Such stories can allow students to see familiar ideas from a completely different perspective. Storytelling teaches students to look for new perspectives by distancing themselves "from the familiar ... seeing something for the very first time" (Pinar et al., 2008, p. 415). Butler (2005) says, "It is only in dispossession that I can and do give any account of myself" (p. 37). Discourse with oneself and then with others allows us to form new ideas. This takes a willingness to forget all old and preconceived notions before pressing onward to new territory.

Telling stories employs multiple relationships. Connelly and Clandinin (2000) sum up these relationships by saying, "participants are in relation, and we as researchers are in relation to participants" (p. 189). Participants and researchers may be experiencing similar phenomena, which brings them common ground in learning situations in education. This excerpt from Schubert (2009) shows the connections between stories, ourselves, others, and education:

Stories

My stories

Whose stories?

Stories of unknown Others

...

In mutual pedagogic relations

Together,

Seeing education

Not as mere schooling

Rather as gathering to explore together ...
Improvises who [we are]
Experiencing, knowing, and doing
Needing and overcoming
Being and becoming
Sharing and contributing
Wandering and wondering ...
(Schubert, 2009, pp. 229–232)

A storied or narrative inquiry explores this intertwining of relationships and allows us to examine the multiple dimensions present in lives, schools, and curriculum. How will I create stories and use stories effectively in my classroom? Through writing conversations, I can tell multiple stories. Mandy's story, Henrietta's story, and the story of my students interacting with Henrietta Lacks. I began to wonder how the authors of the works on my bookshelf would interact with the story of Henrietta Lacks. Could the authors interact at an imaginary academic conference at Johns Hopkins? Maybe Mandy Hoffen could attend and meet and converse with these scholars. It would be amazing if Henrietta Lacks herself could make an appearance and tell her side of the story!

NOTES

1.   The posthuman condition "implies the merging of humans and machines in order to enhance or improve human capabilities" (Weaver, 2010, p. 11). The life of Henrietta's cells has been enhanced by technologies in the field of biotechnology. With these technologies, her cells have become immortal. Posthumanism also calls for "forms of democratic education, curriculum and pedagogy that deconstruct the common sense, taken-for-granted naturalness of humanism, not from an antihumanist perspective, but as a movement beyond the limits and contradictions of the humanist project while still maintaining the modernist and humanists' projects of rights, justice, equity, and freedom" (Carlson, 2015, p. x).
2.   Benchmark assessments are given to students each grading period to see if they have mastered the standards being presented during that grading period. The benchmark does not count for a grade. The benchmarks in my biology classroom are given in the school computer labs. Benchmarks are also used for teacher assessment.
3.   Looking at curriculum through a postmodern lens will involve using many perspectives in lieu of linear standardized approach. See W. E. Doll (1993).

CHAPTER 2

MEET HENRIETTA

PRESCRIPT: PURPOSE FOR CONSPIRACY:
THE OTHER, NEW PERSPECTIVES, AND NEW STORIES

Henrietta Lacks was one of 10,000 women to die of cervical cancer in 1951. A young mother of five, she died a very painful death from an extremely aggressive cervical cancer. The "tumor was biologically unique … peculiar appearance and a neoplasm which was to prove quite resistant to radiation, for Henrietta Lacks was dead within 8 months with widespread disease" (Jones et al., 1971, p. 945). Henrietta's body was moved from the public ward, where she died, to the morgue where her "body lay on a stainless-steel table in the cavernous basement morgue. For research purposes, samples were cut from Henrietta's body: bladder, bowel, uterus, kidney, vagina, ovary, appendix, liver, heart, and lungs" (Skloot, 2010, p. 90). Her abdomen was black and charred from radium treatments, and "tumors the size of baseballs had nearly replaced her kidneys, bladder and uterus … her other organs were so covered in small white tumors it looked as if someone had filled her with pearls" (p. 90). While she lay on the autopsy table, "HeLa [cells] with a generation time of about 24 hours, if allowed to grow uninhibited under optimal cultural conditions" were taking over the world (Jones et al., 1971, p. 947).

Henrietta was taken from that cold table in the morgue and carried back to Virginia to be buried in an unmarked grave. Meanwhile her lauded cells went on to change the history of medicine.

Mandy Hoffen and a Conspiracy to Resurrect Life and Social Justice in Science Curriculum With Henrietta Lacks: A Play, pp. 21–47
Copyright © 2021 by Information Age Publishing
21

HeLa cells make a great science story, but what does Henrietta have to do with my ambitions of creating spaces, weaving stories, building relationships, and creating a conspiracy to resurrect life in my classroom? The bottom line is Henrietta was marginalized in more than one way. She was Black and living in times of segregation. She was a Black woman during a time in history where women had few rights. She was in a public ward at the mercy of the doctors and nurses taking care of her. I could consider Henrietta as "other" because she is different from me. Then I think, at times aren't we all "Other" depending on the lens of examination? I look at the students in my classroom. I see students marginalized, as a standardized curriculum is shoved down their throats at breakneck speed without regard to their race, gender, class, or individuality. I cannot ignore the differences between myself and my students. Barone (2000) says that, "indeed, there must reside within the soul of any true educator a respect for those whom he would educate" (p. 18); a respect for all regardless of difference.

ACT II

The conversation portions of Act II take place at The Johns Hopkins Hospital, a nonprofit academic medical and research center in Baltimore, Maryland. To bring science and scientists together in a historical and collaborative setting, I created a fictional academic conference. The World Alliance of Science and Science Education Symposium is being held February 3–8, 2016. Sessions to honor the 65th anniversary of HeLa cells are also on the program. The anniversary of the first cells to be grown outside the body is monumental to medical research. This anniversary also marks the day that a woman named Henrietta Lacks had a tumor removed at Johns Hopkins. Her cells were taken during a biopsy of a tumor. The creation of this cell line created a new space. A space between a woman and a human cell line, yet the woman's identity is erased. Mandy is struggling with how to tell Henrietta's story in a place where HeLa cells are important, but Henrietta is not. Mandy is also struggling with teaching students in an environment where test scores are important, but students are not.

In Act II Scene I, the first session of this imaginary conference, a group of authors discussing books that are written as conversations. In Act II Scene II, Mandy Hoffen finds herself in a conversation with an artist about a BioArt display, a form of art featuring biotechnology. Finally, in Act II Scene III, Mandy speaks directly to a playwright about the importance of finding an effective method to deliver a story.

Act II Scene I: Conversation Concerning the Written Conversation

Characters in order of appearance: Mandy Hoffen, graduate student/ teacher; Anthony Parker, graduate student; Christopher Vincent, graduate student; Joan Crenshaw, graduate student; Michelle O'Yang, graduate student.

A preconference session for the World Alliance of Science and Science Education Symposium[1] is being held in the George Peabody Library at Johns Hopkins University. Graduate students are preparing to discuss books written as conversations. *The Immortal Life of Henrietta Lacks* by Rebecca Skloot, *The Cambridge Quintet* by John Casti, *A Curriculum of Imagination in an Age of Standardization* by Robert Lake, *Troubling the Angels* by Patti Lather, and *A River Forever Flowing* by Ming Fang He will be highlighted. Mandy Hoffen has her copy of *The Immortal Life of Henrietta Lacks* tucked safely in her bag. She is leading this discussion. A group of graduate students have gathered around a table in a small conference room.

Mandy Hoffen: Welcome everyone. It is an honor to be here with you. I am grateful for this opportunity to linger with you and find our connections with the works we are discussing today. We can tell a little about ourselves as we take our turn. I have become very interested in writing imaginary conversations between authors of multiple disciplines to address current problems in education. All the books we are discussing today have conversations as part of the narrative, some fictional and based on imagination, and some classified as nonfiction. Maybe later we can address possible issues with calling any conversation nonfiction. As writers and researchers, we all have lots of options and lots of opinions when it comes to presenting narrative. The book I want to bring for discussion today is *The Immortal Life of Henrietta Lacks*, written by Rebecca Skloot. Skloot wrote this book following "more than 1,000 hours of interviews with family and friends of Henrietta Lacks, as well as with lawyers, ethicists, scientists, and journalists who've written about the Lacks family" (Skloot, 2010, p. xiii). Skloot includes places where the dialogue between characters tells the story. Skloot worked very hard to "capture the language with which each person spoke and wrote dialogue" so that the dialogue "appears in native dialects; passages from diaries and other personal writings are quoted exactly as written" (Skloot, 2010, p. xiii). Henrietta's family and other participants in the interview felt that this is a very honest way of writing. For those of you who may not be familiar with Skloot's book, she writes about Henrietta Lacks' life, death, and ultimately her immortality. In this kind of story, I think the biggest challenge would be having to close gaps in the story. Some gaps she chose to close, others she left open. When telling a story, "we have an obligation to each other to respect our stories and learn from these stories" (Coles, 1989, p. 30). These stories require a "teller's thoughtfulness, canniness, sensitivity and talent" (Coles, 1997, p. 93). The product that "emerges, if it is done successfully is a kind of truth ... and penetrates deep within one, that leaps beyond verisimilitude or incisive portrayal, appealing and recognizable characterization, and lands on a terrain where the cognitive, the emotional, the reflective, and the moral live side by side" (p. 93). The books represented today are similar due to their creative formats ... narrative and conversations that occur in one place and time, negating actual history or basic chronological order. We have fiction books and nonfiction books in this group. Let's begin by talking about the specific format in the books you bring to the table for discussion today. How about we start on the right and go around.

Anthony Parker: It is very nice to be here. I will be sharing John Casti's work *The Cambridge Quintet*. Casti is "attempting to convey in a fictional setting the intellectual and cognitive issues confronting human beings involved in shaping the science and technology of their future" (Casti 1998, p. xi). By adding an element of fiction, he can "imagine how the world we live in today was shaped by decisions of the past, and how the decisions we make today impinge upon the world of the future" (p. xii). Make sure you understand that participants in this "fictional account of the hypothetical—but possible—gathering presented here will on occasion see the participants making statements in ways that depart from what we might imagine they would have said on the basis of their published works" (p. xii). "For the sake of exposition, [Casti has] moved several conceptual themes in AI [artificial intelligence], from their actual time in the post-1950 decades back to the period of this dinner" (p. xiii). So, the work shifts certain time elements.

Hoffen: I thought the most interesting part of The *Cambridge Quintet* was the setting.

Parker: I agree. The setting for this book is a fictional dinner party. I guess for Casti, that seemed like a natural choice. The chapters are creatively divided up by the names of dinner courses. Casti explores "a conflict of ideas pitting Ludwig Wittgenstein and Alan Turing on opposite sides of the issue: Can a machine think?" (Casti, 1998, p. xii). In this book the ideas presented are "the message" (p. xii).

Hoffen: I am so happy for an opportunity to discuss this book. A colleague suggested I read it. Honestly, it was just a great read. I wondered why Casti did not just write a popular science piece. Maybe focus on today's issues, and simply use a historical backdrop.

Parker: That might have been easier for him. If writing a popular science piece, he "would have been limited to what is known about the motives and thoughts of the people involved" (Casti, 1998, p. xii). If he had chosen science fiction or a general novel, "then the story would have had to adhere to the principles and conventions of those genres, concentrating on the development and change of the world views of the book's characters through the resolution of conflicts" (p. xii). Casti chose science fiction. He "wanted to present a lively and comprehensible exposition of the intellectual and emotional uncertainties involved in shaping the future of human knowledge" (p. xii). "Science fiction has a mission to try to imagine how the world we live in today was shaped by decisions of the past, and how the decisions we make today impinges upon the world of the future" (p. xii).

Hoffen: I am wondering how he handled the legalities of writing words for characters that are no longer living? I am sorry for my interruption. I have so many questions today. We will come back to that later. Thanks, Anthony. Christopher is next.

Christopher Vincent: Thanks, Mandy. It is all about imagination for me. I have been studying Maxine Greene extensively for several years. A professor encouraged me to read Robert Lake's dissertation. Lake wrote imaginary conversations between Maxine Green and Paulo Freire. Lake (2008) writes in *A Curriculum of Imagination in an era of Standardization* that dialog "adds a humanizing dimension and makes [the narrative] come to life in clear and understandable terms" (p. 113). In Lake's dissertation he explains that "the conversations were arranged in a way that helped [him] understand the role of imagination in developing [his] own personal agency in removing barriers to social justice, first within [himself], and then in the lives of those whom [he teaches]" (p. 113).

Hoffen: Lake's goal sounds familiar. He too created a fictional dialogue, bringing together people to converse about issues today.

Vincent: "The works of Greene and Freire epitomize two aspects of imagination: artistic and critical, that are crucial to understanding what exactly is meant by [imagination] when it is applied to the context of education" (Lake, 2008, p. 113); just as "the fusing together of film is often done in juxtaposition in ways that create greater depth and width to the representation of images" (p. 113). Lake (2008) was able to create "a montage of quotes from Green and Freire, in writing" (p. 113). This arrangement of quotes "provided a way to internalize a discursive interaction between the creative and critical aspects of imagination by the blending of [Greene and Freire's] written voices with [Lake's] own inner voice of personal musing" (p. 113). Writing such a dialogue is like an artist blending colors together on a palette, and then placing the individual and blended colors on a canvas. That sounds nice doesn't it, like blending colors on a canvas? Very poetic if I say so myself. Writing what you think others say involves a lot of risk. Personally, I had to constantly wonder if the words Lake was writing for Maxine Green and Paulo Freire represented what they would say. I would be afraid that I would misrepresent the writer's intentions.

Hoffen: That is a good point. How much license as writers can we take when writing around direct quotes? Writing is an art form. Rebecca Skloot moved back and forth between two stories in *The Immortal Life.* She had to focus on chronology: "Dates for scientific research refer to when the research was conducted" (Skloot, 2010, p. xiv). Was time or chronology a factor in Lake's conversations?

Vincent: Anthony alluded to this earlier: Time can be flexible in some works of this type. Skloot's book, on the other hand, did involve actual scientific discoveries. Her book is more of a factual/nonfiction account ... as much as nonfiction can be factual. Staying true to actual time was necessary in *The Immortal Life of Henrietta Lacks.* Lake brings his discussion of imagination between the two participants to modern day, to solve a modern problem, making time a less important aspect

Hoffen: Thanks, Christopher. This next book is another one of my favorites.

Joan Crenshaw: I look at the books you brought with you today, and I must say mine appears to be very different. When I say different, I am speaking about the actual layout of the text on the pages. *Troubling the Angels* (Lather & Smithies, 1997) has two dialogues running across the pages. Patti Lather wrote *Troubling the Angels* with Chris Smithies. The book is rich in dialogue, but the dialogue is actual, and in real time. Unlike the works by Bob Lake and John Casti, the conversations that are presented are those of the actual participants in the women's HIV support group featured: "Across the bottom of much of the book is a continuously running commentary by Smithies [and Lather], the coresearchers, regarding [their] experiences in telling the women's stories that moves between autobiography and academic 'Big Talk' about research methods and theoretical frameworks" (p. xvii). Both stories are told simultaneously, like Rebecca Skloot's account of Henrietta's story. Skloot's story flips back and forth between chapters. Lather and Smithies' work physically flips on the page. Top of the page—the women in the HIV support group. Bottom of the page—the authors' contemplations concerning the women and the horrible disease AIDS. I think the term lingering can be important here. This multiple perspective approach allows the reader to linger in multiple settings to contemplate Henrietta's story. Its more than a flipping back and forth, true lingering will allow the reader to experience the story in multiple settings through multiple sets of characters.

Hoffen: I had to read the two sections separately. I read what the women were saying and then read the author commentary on the bottom. Thanks so much, Joan. Michelle is next. She traveled the farthest today. She comes to us by way of China.

Michelle O'Yang: There are a few connections that I think bear mentioning with the book I bring to the table today and the others discussed. I will be discussing Ming Fang He's *A River Forever Flowing*. Ming Fang He was the chair for Bob Lake's dissertation committee. I can really relate to Ming Fang He's work because my life and work has put me in situations of being in-between. She writes about her experiences of living between cultures. I too find myself situated somewhere between East and West ... as I lived and worked in China, then here in the United States. When I went back to visit my family in China after being a student in the United States for 2 years, I no longer seemed to be the same person I had been when I left. *A River Forever Flowing* is a "cross-cultural narrative approach" that "creates new ways to think about, talk about, and write about cross-cultural lives, cross-cultural identities, and their relevance to multicultural education" (He, 2003, p. xvii). He (2003) shares the stories of "three Chinese women, teachers as they moved back and forth between China and Canada" (pp. xvii–xviii). The teachers meet and talk. Her book records the conversations. In order to protect the identities of herself and the other teachers, she created composite characters, each sharing portions of each other's stories.

Hoffen: That is a very creative way to keep a character's identity confidential. My work has an opposite issue. I want to create characters from my bookshelf. I want to use their work in creative conversations, but I want to use their real names. Most of what these characters will say will be direct quotes from their own work. In order to create an imaginative story line ... I sometimes need to add words—my words. There are ups and downs to writing conversations. You are not just writing your ideas; one must represent the ideas of others ... not just in words ... but in a manner that reflects their personal beliefs, values, and even their personalities. I think it is a complex technique used to tell a story. Composite characters might be less complicated.

Parker: So, you are wondering how much license you have because you are writing a fictional conversation?

Hoffen: Exactly, I began experimenting in graduate school with writing conversations. As a graduate student I was constantly reading different authors suggested by my professors. The reading was very interdisciplinary. In a created discussion format, these authors could come together and converse about a topic in a way that could not happen otherwise. My first dialogue was written to highlight ideas about science and the scientific method using authors: Herman Melville, Henry David Thoreau, Ralph Waldo Emerson, Walt Whitman, and Alexander von Humboldt. I created a fictional classroom. This group of authors became the students in this fictional classroom ...

Hermie Melville, Ralphie Emerson ... you get the idea. In the story their revolutionary ideas were driving one of their teachers crazy. This teacher was a very close-minded individual whose ideas represent what education should not be. The other teacher was open to the ideas of these young scholars. My dissertation chair suggested I read the Casti book. With imagination—any gathering is possible. I found that adding fictional settings to the conversations gave another dimension to the conversation. Another professor asked me to write a literature review for a dissertation-writing class in a creative manner. I honestly thought it was a little odd to write a literature review in the form of a conversation. Once I started ... it seemed to make perfect sense. I could put on paper the conversations going on in my head. Before long, I was on a quest to read books written in conversation. I had always been a fan of books written in interview form ... such as *Conversations on Science, Culture and Time* (Serres & Latour, 1990/1996). In this book, Bruno Latour interviews Michel Serres. The format allows for the participants' autobiography, history, culture, and relationship to science to come together. A straight narrative could not capture the multiple meanings. Fortunately, my professors continued to suggest other works dealing with conversation. Each one that I have read has added a piece to the puzzle, as I explore and begin writing my research. *The Immortal Life of Henrietta Lacks* has become the foundation for my work. I am asking the question: Can Henrietta herself enter conversations with scientists, science historians, science philosophers, professors, and students? Can Henrietta's story become as much a part of science as her HeLa cells? Her cells have been the object of conversations in science for many years. It was not until *The Immortal Life* that so many people began to include the woman behind the cells in their conversations. The story of her cells is an obvious addition to science class. But what about the woman? The real woman, who suffered and died right here at Johns Hopkins. I often wonder what kind of conversation philosophers such as Michel Serres, Donna Haraway, and Bruno Latour would have about Henrietta Lacks, or even with Henrietta Lacks. What would they think about her being part of the curriculum in science classrooms? As I form all these questions, I wonder how I can go about formulating an answer, or a response for each one. The idea came to me that conversations between selected authors could help me formulate an answer. I knew also that the answer would only come if science could intersect with literature. Weaving conversations was like creating art. Now it seemed everything was getting more complicated. Conversations, art, science, literature, me, Henrietta …

Vincent: The best part about these conversations is that you are using your imagination. In the conversation you can bring many ideas to the discussion, with no barriers of time. I like the weaving metaphor. Where do you think your conversation will take place?

Hoffen: I honestly have not thought about that, Chris. I am thinking that at some point some of the conversation will have to occur at my school … maybe even with my students. Going back to what you said earlier, I am extremely worried about making sure I represent the ideas of each writer that I invite to the conversation. This will not be an easy task.

Crenshaw: Mandy, I seem to understand that you are writing conversations about Henrietta and her story, and then also writing about Henrietta and your story. Don't worry, as you write, the setting and format for your conversations will come to you. You do have some wonderful examples that have been presented today. If everyone will bear with me just a moment, I want to ask our student a very necessary question. Mandy, you are enthusiastic about Henrietta's story, but so are the millions of people that read and bought the book that has occupied the best seller list for more than a few years. Don't think I am making light of your enthusiasm. Can you tell us what you hope to accomplish by telling her story? It is obvious that you want to instigate change … but be specific … I want to know what kind of change can take place because you have a conversation about Henrietta Lacks.

Hoffen: I have had to defend my work on multiple occasions. Presenting at conferences has also helped that process. I do want to change the world. I know that in order to change the world, I must find a platform. My platform for this project is my work as a practicing teacher teaching a standardized curriculum. I know I must create a space where transformation and learning can take place. Let me try to give you an example of this space using the Michel Serres "river" metaphor from his book *The Troubadour of Knowledge* to explain learning (Serres, 1991/ 1997, p. 7). My students and I are standing on a riverbank. We must jump in the water and completely leave the shore.

According to Michel Serres it is in the crossing of the river that we learn:

> In crossing the river, in delivering itself completely naked to belonging to the opposite shore, it has just learned a third thing. The other side, new customs, a new language certainly. But above all, it has just discovered learning in this blank middle that has no direction from which to find all directions. (Serres, 1991/1997, p. 7)

I want to focus on what goes on in the middle of the river. In all honesty, we may never make it to the other side. While we are in the raging waters of this twisting, winding, raging river, we lose all sense of direction. We have no idea how to get to a specific place on the other shore.

When I share the story of Henrietta with my students, I am giving students a space where they can connect to Henrietta's story. A space where students can see that science is more than the standardized curriculum that must be taught and tested. I wonder if my students can connect to Henrietta on a personal level. I also hope they will see that they must share Henrietta's story because she cannot speak for herself. As they begin sharing her story, I want them to realize that Henrietta is not the only person in the world who no longer has a voice to bring about change. My students need to know that they must use their own voices and become activists for change in our world.

Vincent: You know, writing itself is a research method. I wish we had time for that discussion.

Crenshaw: *In Troubling the Angels*, Patti Lather writes how difficult it is when someone does not want to be included in a literary work. The researchers wanted to add one woman's story to the book, but the woman was afraid of being exposed. She said her story was "too personal ... I don't want people to gawk at me ... now I know how the movie stars feel to have everyone looking into their business" (Lather & Smithies, 1997, p. 200). People's privacy must be respected. Lather says that with each participant she had to "face the possibility that [some participants] will not feel comfortable with being included" (p.200), She says she felt guilty for her attempts to cajole these women "against [their] own best interests, in spite of [herself] and [her] deepening understanding of and gratitude for what it means for these women to put their lives on such public display" (p. 200).

Hoffen: When you think about it like that, that is a lot of responsibility. Skloot took the responsibility very seriously as she worked with Henrietta's family. Ascribing words in a fictional setting is a different issue ... or is it? I think those individuals need to be respected in the same way. I think if we are writing about individuals who are no longer living, they still must be treated with the same respect. In Casti's book, most of his participants are dead. In Dr. He's book, the names of participants needed to be protected because of censorship that remains present in China. Should He not have written this work?

Is our time up already? We better stop here so everyone can get to their next session. It was nice to meet you all. I hope we can meet up and continue this conversation.

Hoffen's head is spinning. She has a lot of new ideas concerning her conversations. As she walks to the next session she has selected, she finds herself in the middle of another conversation.

Act II Scene II: Art and Science Converge

Characters in order of appearance: David Blades*, professor; John Mighton*, professor; Peter Anderson, professor; C.P. Snow*, professor; Adele Senior*, professor; and Mandy Hoffen, graduate student/teacher.

A small group of people are gathered in the lobby of Johns Hopkins Hospital and Research center, attending a meeting being held in conjunction with an anniversary Cell-e-brating of HeLa cells—the first human cells to be grown in culture. A group of conference attendees are contemplating a display.

The display is called "HeLa." It was created by a Swiss artist, Pierre-Philippe Freymond (2006).[2] The display "consists of an incubator containing living HeLa cells in a nutrient medium and an inverted binocular phase contrast microscope. In addition to this scientific equipment, Freymond (2006) includes a black and white photograph of Henrietta Lacks, a light box, a neon light displaying the possessive noun Henrietta's, a mural and an A5 booklet" (Senior, 2011, p. 524). The group is taking turns observing the cells displayed under the microscope. The room is dark. The only light is coming from the neon pink sign and the microscope's light source.

Blades: Is it art? Is it science?

Mighton: Well, if I remember from my freshman biology class ... the microscope slide does sit on a stage?

Anderson: Ha, a little ninth grade humor there …

Snow: I have envisioned science meeting art. This seems to be a bit of both.

A woman approaches the group, hearing pieces of the conversation.

Senior: It is BioArt—"an artistic attempt to negotiate the history of HeLa and Henrietta Lacks" (Senior, 2011, p. 514). Rebecca Skloot's approach involved "recording the life and death of Henrietta Lacks" (p. 514). Have you all read *The Immortal Life of Henrietta Lacks?*

Blades: No, but maybe this is something I need to check out. *The others were shaking their heads no.* You called this display BioArt?

Snow, Mighton, and Anderson all repeat the word "BioArt" simultaneously, each with their own individual puzzled look.

Senior: BioArt is the practice of using biological materials such as cells, tissues, DNA, in artwork. "The increased use of biological material, such as HeLa cells, in art practices over the past decade has given rise to a fleshlike 'object' which is reminiscent of performances' challenge to the traditional archive" (Senior, 2011, p. 511). "Biotechnological tools and practices prompt new (and reiterate existing) challenges for a politics of visibility and an ethics of responsibility to the other" (p. 512). "Artists working in this area utilize the tools of biotechnology and, in doing so, often include a living or biological component in their practice. Biotechnological techniques, such as tissue culture or genetic engineering, and living or biological materials such as cells, tissue, bacteria and viruses, are frequently employed as artistic media in BioArt practice. Artists and art collectives working in this field have turned to the science lab and the tools and techniques of the life sciences for a number of reasons with diverse intentions" (p. 515).

Snow: Amazing. Definitely the merging of two cultures—yet more than I ever imagined possible. Let me explain. Years ago, "By training I was a scientist: by vocation I was a writer … there have been plenty of days when I have spent the working hours with scientists and then gone off at night with some literary colleagues" (Snow, 1959/2013, p. 1). Long before I made a sketch, on paper, of this phenomenon I began calling it "two cultures" (p. 1).

Blades: Ah, would you happen to be *the* C.P. Snow? I think we have all read your lecture on the "Two Cultures" (Snow, 1959/2013, p. 1).

Snow: Yes, that would be me. Have we made advances in this area? When I gave my lecture, a decade or so ago, I insisted that "Western society is increasingly being split into two polar groups: At one pole we have the literary intellectuals, who incidentally while no one was looking took to referring to themselves as intellectuals as though there were no others" (Snow, 1959/2013, p. 4). You could describe this as intellectuals at one end of the spectrum, and scientists at the other. "Between the two a gulf of mutual incomprehension—sometimes (particularly among the young) hostility and dislike, but most of all lack of understanding. They had a curious distorted image of each other" (pp. 4–5). The attitudes of both groups of individuals keep them from being able to find common ground!

Are the same stereotypes I wrote about still in existence? Long ago, "nonscientists tended to think of scientists as brash and boastful … shallowly optimistic, unaware of man's condition" (Snow, 1959/2013, p. 5). On the other hand, the scientists believe that the "literary intellectuals are totally lacking in foresight, peculiarly unconcerned with their brother men, in a deep sense anti-intellectual, anxious to restrict both art and thought to the existential moment." (p. 6). My life was crazy:

> I felt I was moving among two groups—comparable in intelligence, identical in race, not grossly different in social origin, earning about the same incomes, who had almost ceased to communicate at all, who in intellectual, moral and psychological climate had so little in common that instead of going from Burlington House or South Kensington to Chelsea, one might have crossed an ocean. (Snow, 1959/2013, pp. 2–3)

As a spectator in the middle, in the in-between spaces, I declared "the number 2 is a very dangerous number: That is why the dialectic is a dangerous process. Attempts to divide anything into two ought to be regarded with much suspicion" (pp. 9–10). "The clashing point of two subjects, two disciplines, and two cultures—of two galaxies, so far as that goes, ought to produce creative chances. In the history of mental activity that has been where some of the breakthroughs came" (p. 17).

Senior: The clashing of biology and art can lead to such break-throughs. The artist considers this display a performance. "These performances work to challenge the discourse of expertise and fear within which a discussion of these issues is ordinarily couched and encountered in the public realm" (Senior, 2011, p. 515). Without the clashing of the two supposedly historical polar entities you call "two cultures," the break-through is nonexistent (Snow, 1959/2013, p. 2).

Anderson: So, a breakthrough somewhere between biology and art—hence BioArt. I see similarities here. The display allows us to entertain the idea that Henrietta's life impacted the discipline of science, not just her cells, but Henrietta as a person. What is the significance of the neon sign? I know they were popular in the time that Henrietta lived.

Senior: The original display is separated from the rest of the gallery. The sign can be seen from all over the gallery, but one must come closer to observe Henrietta's photo and cells. This arrangement "reads as the transition from the public to the private world of the voyeur who secretly observes the private life of an unsuspecting other" (Senior, 2011, p. 525). The idea concerning Henrietta's cells being "removed and used without her knowledge takes on a more immediate and potentially uncomfortable resonance in an installation in which HeLa cells are magnified and Lacks's photograph is displayed alongside them in an intimate but nonetheless public exhibition" (p. 525). The artist had a very deliberate purpose in using a neon sign. "The sign initiates a nostalgic reference to an imaginary 1950's diner called Henrietta's" but its purpose is twofold (p. 525). The neon sign "inevitably attributes ownership to Lacks by employing a possessive noun" (p. 26). "What exactly is owned by Henrietta is for the spectator to decide (is it Henrietta's image, Henrietta's cell line or Henrietta's story?)" (p. 526). The "installation instead creates a space not only to imagine a different kind of history but to imagine a different kind of archival practice" ... in order to secure memory (pp. 526–527). Memory of the cells, memory of the history, and memory of the woman—Henrietta Lacks. Sorry, I did not mean to get carried away. I am a graduate student, and I get a little carried away when I have an audience willing to listen to my work.

Blades: What was your name?

Senior: Adele Senior. I have followed the rise of BioArt as a form of expression.

> As the use of biological material in art practice necessarily raises ethical questions, perhaps the most ethical response is to acknowledge the otherness in which BioArt practices participate within the acts of receiving as well as producing BioArt. Indeed, despite repeated attempts in the academy to give Bio-Art to disappearance, these bodily responses (bone against flesh, flesh against bone) will remain and, like HeLa and Henrietta, will (and perhaps should) always return to haunt us. (Senior, 2011, p. 529).

Hoffen: Henrietta haunting us ... now there is an idea. In Skloot's book, her family makes mention of her making her presence known from the grave.

Blades: BioArt—very interesting work. This work effectively demonstrates that efforts are being made to bridge the continuing binary of your "two cultures," Dr. Snow (Snow, 1959/2013, p. 2).

Hoffen: I heard about a new initiative for science, technology, engineering, art, and mathematics to all be taught in a connected interdisciplinary fashion. The program is called STEAM.[3] Unfortunately, because it is only offered to upper-level students in my district, it appears that this program is only designed to further the current segregation in education. These options are reserved for the gifted and talented students. That is the part that gives me the most trouble. I think all students deserve to see how what they are learning connects to their lives and the world in which they live. Students need their own in-between space to contemplate what they are taught and how it relates to them personally. My district is implementing STEM ... not STEAM ... why are the poets and artists of the world neglected? Sorry, I also tend to get carried away when I am speaking about my own graduate work. It is nice to meet you, Adele. My research includes work with Henrietta Lacks. BioArt seems to be creating bridges, but in our schools, I feel that your "two cultures," Dr. Snow, are alive and well (Snow, 1959/2013, p. 2). Subjects are taught according to lists of objectives. Most of these lists for science classes involve students memorizing facts on mandatory standardized tests and benchmarks. Standardization has extended the chasms separating subject matter such as science, humanities, math, and history, into isolated lists of objectives that teachers must post on their walls.

Blades: Funny you mention STEM, here. I have written a play about STEM, and its impact on education. Think of it as science education meets the stage. Two of my colleagues and I performed the play at a recent conference.

Hoffen: Wow, I bet it was "really nice to see something other than a PowerPoint presentation" at an educational conference (Ferguson, 2014, p. 1). What was the main idea behind the play?

Blades: "Internationally, STEM has become a slogan for organizing new discourses and practices in science education" (Weinstein et al., 2014).

Hoffen: STEM certainly is the acronym getting the most attention in the district where I teach.

Blades: Our one-act play "explores the role STEM could play in stimulating the invention, deployment, and development of alternate sources of electrical power…Through a rehearsed dialogue presented in the form of a one-act play we use the discourse of alternative power to reveal possibilities for reimagining STEM in new ways" (Weinstein et al., 2014). The dialogue on stage "demonstrates the possibility of alternative pedagogies to resist the dominating discourses of national competitiveness. Our hopeful conversation thus suggests new roles for school science education outside and against the prevailing discourse of neoliberalism" (Weinstein et al., 2014).

Hoffen: Wow, I had not thought of STEM in that way. Once a word is introduced, it finds root, and no one even bothers to question what is going on. I am still a little stunned about you presenting a play at a conference. Most academic papers are not written in script form.

Mighton: I think it is very unfortunate that something like STEM is set aside for an exclusive group of students. Unfortunate indeed. Like BioArt, having this discourse about STEM on stage creates a place where a necessary discussion can take place. Like the two cultures of science.

Anderson: Just in the last decade or so, several plays have appeared—plays which deliberately seek to reunite the "two cultures," bringing modern physics and math directly to the stage (Innes, 2002, p. 21). John, your plays have played a role in closing the gap between two cultures. Dr. Snow, John Mighton here has life experience like yours. He holds doctorate degrees in philosophy and mathematics. He is also a professional playwright.

Mighton: Nice to meet you all. Most of my plays do bring topics of physics and math to the stage.

Anderson: Needless to say, I am a huge fan. These plays eliminate the deep and disastrous chasm between Dr. Snow's (1959/2013) "two cultures" (p. 2). When you see science on the stage, something interesting happens:

> Humanistic values are applied to science, while at the same time scientific principles are incorporated into artistic structure. And beyond that contemporary science, whether quantum mechanics, complementarity and the uncertainty principle, or fractal geometry, thermodynamics, and chaos theory are asserted as positive forces. (Innes, 2002, p. 26)

Mighton: Most of my plays deal with mathematics and physics. I want to see individuals imagine themselves in other "possible worlds" (Mighton, 1988, p. 23). We "exist in an infinite number of possible worlds. In one world I'm talking to you right now, but your arm is a little to the left, in another world you're interested in that man over there with the glasses" (p. 23). "Creatures like us that can anticipate possible futures and make contingency plans have an evolutionary advantage. We would be foolish not to use our imaginations, not to investigate every possible fact" every nook and cranny ... every in-between place (p. 26). "The possibilities swarm" (p. 61). "I also know that what you have is always relative to what you can imagine" (p. 51).

Hoffen: Hmm. Use your imagination and make your presentation on stage ... very untextbook. What a great way to represent your ideas. With a stage, ideas of the past, present, and future can all meet in one place at one time and address specific issues of the present. What about those who feel that education should be served up in nice isolated portions—all disciplines being taught as separate entities?

Mighton: Some may not know much about math and chaos theory. It is sad that "chaos has a bad name" (Mighton, 1987, p. 21). However, "All-natural processes depend on [chaos]. Chaos introduces novelty" (p. 21).

Hoffen: Curriculum theorist William Doll (1993) uses chaos theory to talk about education. He uses a graphic called the "owl's eyes" as a metaphor to explain possibilities in education (p. 92). In popular terms, a linear system is exactly equal to the sum of its parts, whereas a nonlinear system can be more than the sum of its parts. This means that in order to study and understand the behavior of a nonlinear system one needs in principle to study the system as a whole and not just its parts in isolation. Education should be viewed with a similar perspective. Especially when we start talking about linear, monodimensional lists of standards for teachers to teach and students to learn.

It has been said that if the universe is an elephant, then linear theory can only be used to describe the last molecule in the tail of the elephant and chaos theory must be used to understand the rest. Or, in other words, almost all interesting real-world systems are described by nonlinear systems. Novelty is what we need! I can imagine a world with Henrietta Lacks, not just her HeLa cells. I can imagine science on the stage with philosophers, historians of science, university scholars, public school teachers, students—geeks, artists, football players—all talking about Henrietta Lacks and her amazing cells. All united and knowing that there must be more than a monocular view of the worlds and its issues of social justice. I am not sure if I am thinking of a haunting here. I like Barone's use of the word "conspiracy" (Barone, 1990, p. 313). He is not using the word conspiracy to describe something "evil" or "treacherous" but defining it as a "profoundly ethical and moral undertaking" (p. 313). He goes on to say a conspiracy is "a conversation about the relationship between present and future worlds" (p. 313). I want the world to meet Henrietta up close and personal. Meet her, take a walk, converse with her. I want to hear her words, hear about her cells, her experiences, from her in person. I also want each of my students to have a voice in their own education … their own learning. Join with others to have a conversation about "present and future worlds" (Barone, 1990, p. 313). I hate to stop here but, we better hurry or we are going to be late for the next session.

Everyone leaves the exhibit hall rather abruptly. The stage darkens with only the work of art remaining lit. A lightning bolt originating in the neon sign strikes the culture dish. The culture of HeLa cells begins to bubble violently. The cells spill over on the table, then to the floor. Miraculously, the puddle is morphing into a humanlike figure from the floor up. The audience only seeing a silhouette watches eagerly as the figure walks slowly to the wax statue of Henrietta Lacks on display by the entryway of the exhibit. The figure removes the clothing and shoes from the life-size mannequin and dresses in the darkness. She steps into the light, straightens her hair, and proceeds into the lobby of Johns Hopkins Hospital and Research Center.

Act II Scene III: Science and the Theater

Characters in order of appearance: Mandy Hoffen, graduate student/teacher; Alan Brody*, university professor of theater; small audience.

The next morning, conference organizers report the theft of the mannequin's clothing to campus security and the local authorities. The artist is working to clean up the display so viewing can continue, as scientists and science educators gather to cell-e-brate the anniversary of those immortal HeLa cells and their donor, Henrietta Lacks.

Mind still spinning, Mandy sits down with her catalogue that describes the many conference sessions—the catalogue is the size of a New York City phone book. She finds herself scanning for anything that has to do with playwriting. Is playwriting a possibility for educational theorizing? She needed more information about that play on STEM. Could she write a play about Henrietta? How would she write such a play? Could she bring science to the stage, like John Mighton? On the verge of giving up, she sees a session entitled Operation Epsilon: Science, History, and Theatrical Narrative with Alan Brody from MIT. The session begins in less than an hour. She jumps up to make her way to the room, looking puzzled as she walks by a bare figure ... a bare figure that had been dressed in a very stylish suit the previous day. Wig still intact ... shoes absent. What horrible person would steal the clothes off Henrietta's back? First her cells ... now her clothes and shoes.

Hoffen walks into the session still bewildered by the bare mannequin. For some reason she walks right up to the front row and sits down. She glances at the gentleman standing in the front of the room at the podium. He smiles and moves the podium to the wall. He pulls up a chair in front of the audience.

Brody: (*Looking at Mandy Hoffen, he speaks quietly.*) I think it is easier to converse with people if you sit with them ... and not stand over them. (*He then looks across the audience to address the group.*) Good morning. I am Alan Brody, professor of theater from

MIT. My session today deals with Operation Epsilon, and the presentation of science on stage. Welcome, I look forward to our discussion. Before we launch what will hopefully be a fruitful discussion for all of us, I am going to read a brief portion of my paper:

> In 1945, shortly after VE [Victory in Europe] day, the Anglo American forces rounded up ten renowned [German] nuclear scientists and interned them at Farm Hall, an estate near Cambridge, England. All the rooms on the estate had been bugged. The conversations of the scientists were recorded on wax discs and translated. Information regarding both the scientists' research and anything else that might be of interest to the Anglo American military was sent to Washington and London. The internship lasted from July to January. During that time, America dropped the atomic bomb on Hiroshima and Nagasaki, and Otto Hahn was awarded the Nobel Prize for his discovery of fission. The men's responses to those events are a part of the transcript. The entire operation had the code name Operation Epsilon: The Farm Hall Transcripts, and in the states as Hitler's Uranium Club. (Brody, 2011, p. 253)

MIT scientists and playwrights began meeting over wine and cheese to examine the history involved in the Farm Hall transcripts as well as to "discuss any random thoughts about science and theater" (Brody, Personal communication.) I saw the story of Farm Hall as my personal "opportunity to explore the range of possibilities in the question of the moral responsibility of scientific research. I was also aware that it would resonate, if only by implication, with another of my concerns, the question of the moral responsibility of the artist" (Brody, 2011, p. 254). "When I talked about the limits of research, I was not talking simply about scientific research, but specifically the more complex moral problem of 'pure research,' that is, theoretical research before the application of technology that claims to be value free" (p. 254).

"I wanted to trace the changes in these men in terms of their moral awakening (or not) about their work for Hitler. From this perspective, I could find the changes that my story would hang on. But how to tell it?" (Brody, 2011, p. 254). My goal was a telling of a story that did not disregard nor take liberties with the transcripts. "I wanted to retain historical accuracy and the immediacy of the transcripts, but I was already moving into that slippery territory of writing history theatri-

cally, the territory where the question of limits and liberties
became vexed" (p. 254).

*This was all sounding too familiar to our graduate student. What had started as a
simple conversation, and a retelling of a story, was evolving into a conversation
involving scholars who had crossed the borders of time to meet in one place. Then
composite characters were evolving to speak and tell the stories from many different
places and time. Before she knew it, she was writing conversations with people who
had risen from the grave, people who had no voice, no choice in whether to be participants in my evolving story. She was speaking on behalf of scholars and university professors. What liberties can be taken? What a scary proposition, writing
words for scholars she had read, and writing words for scholars she knew personally.*

*Brody pauses to get a sip of water, and before Hoffen realizes it the words just
come out.*

Hoffen: How do you do it? How do you weave your story? How do
you set your boundaries? (*People all over the room shake their
heads in disbelief that this graduate student had just interrupted a
famous award-winning playwright from MIT. Brody continues without even a minute display of displeasure.*)

Brody: Well, it is not an easy proposition, no matter what the time,
no matter what the story. "Whenever it comes up, theater
scholars almost automatically point to *Richard III*, Mary *Stuart*,
or *Galileo*, implying that if it was alright for those guys it must
be alright for anyone" (Brody, 2011, p. 255). "What emerged
most clearly for me from those sessions about Farm Hall was
the idea that the critical element has to do with whether the
play, by its very structure, asks us to believe in its historical
accuracy and if the intentions of the play are dependent on it.
The fact that Mary Stuart and Elizabeth I never met, even
though their meeting occurs in one of the great climatic scenes
in 19th-century German theater, seems to be less problematic
for most people than what Brecht does with Galileo" (p. 255).

(*Hoffen scribbles a note to herself to look up Brecht and Galileo.*)

Brody: "Brecht calls his play *The Life of Galileo*, with the implication
that the history resonates with the present. Even with current
ideas about the relativity of historical truth, there are still historical facts to be considered" (Brody, 2011, p. 255). If there is
something historical that is purposely distorted, the intentions

of the play may be undermined: "I am still looking for a paradigm that might locate the boundaries of license" (p. 255).

Hoffen: How do you know how much science to include? I can only imagine conversations about nuclear fission taking place on stage and an audience all dressed up expecting only to be entertained.

Brody: I always ask myself, "How much of it did the audience need in order to follow some of the crucial questions in action?" (Brody, 2011, p. 255). "The technical issue for me was finding a way to work in enough accessible background so that the audience could follow the arguments and so that the characters did not sound as if they were simply talking science" (p. 255).

Hoffen: Is it okay to adjust the facts, so to speak?

Brody: Not everything in a play needs to be considered as truth. "We worry that verity equals verisimilitude...we also worry that "verisimilitude implies verity" (Brody, 2011, p. 257). "That paradox opens up an opportunity for exploring the relationship between life and art, the demands of dramatic truth in the context of fact. It is a rich area of inquiry for the critic" (p. 257). I would suggest that you write in "a realistic mode and try to stay as close to the historical record" as possible. It may take quite a few performances and rewrites to obtain the balance you are looking for.

As far as Mandy Hoffen was concerned, there was no one else in the room, just she and Brody. He had transcripts; she had *The Immortal Life of Henrietta Lacks* and several other tellings of Henrietta's story. Because of Rebecca Skloot, people knew Henrietta's story, but did people know the implications of her story in philosophy, history of science, university classrooms, and high school science classrooms? If the story unfolded on stage, students could witness this firsthand. All participants could create a version of Henrietta's story involving his or her perspective, his or her life experience. Creating a new multiperspective version of Henrietta's story would involve lingering. Lingering in the context of creating bridges between two stories. Lingering in a manner that allows all participants to compare their own story to Henrietta's story

POSTSCRIPT: JUXTAPOSITIONS: ART MEETS SCIENCE
AND A RESEARCHER'S LINGERING

Mandy is experiencing how others share stories —books as conversations, BioArt, and the theater. As she begins to linger and contemplate how she can share Henrietta's story, she is doing so in a space between art and science. C.P. Snow (1959/2013) describes the art-science dichotomy in his lecture titled, *The Two Cultures and the Scientific Revolution*. Snow (1959/2013) believed that "the intellectual life of the whole of Western society is increasingly being split into two polar groups" (p. 4). These two groups are composed of the "literary intellectuals at one pole—and at the other scientists, and as the most representative, the physical scientists. Between the two a gulf of mutual incomprehension—sometimes (particularly among the young) hostility and dislike, but most of all lack of understanding" (p. 4). Snow (1959/2013) explains that "The reasons for the existence of the two cultures are many, deep, and complex, some rooted in social histories, some in personal histories and some in the inner dynamic of the different kinds of mental activity themselves" (Snow, 1959/2013, p. 23). During this time in history:

> The only means of shedding light on some aspect of the physical world was science, and when it came to issues surrounding human beings, social science was the sole illuminating source. For them, the dominant research paradigm was the experiment and was the gold standard for research. The use of statistics was ubiquitous, and the models to be used in doing research had a mechanistic uniformity. (Barone & Eisner, 2012, p. x)

My original impressions of scientific research were embedded deeply into my psyche while working on a master's degree in Biology. This suggests an image of a small wood-framed laboratory on Frenchman's Bay at the Mount Desert Island Biological Laboratory. I collected mountains of quantitative data—precision and duplication were required. This type of data is important. However, not everything in our universe can be reduced to a data point on a graph, or a number in a table. What happens when we begin to look at science differently, maybe through the eyes of artists or poets?

As a doctoral student, with much prior experience of the basic scientific method, my new ideas of how to approach research have been thoroughly challenged. John Weaver introduced me to two writers that I would not have identified as scientists: Ralph Waldo Emerson and Alexander von Humboldt. I had never heard of Humboldt, an explorer and scientist whose history and ideas have been swept under the proverbial rug. We studied Humboldt through the work of Laura Dassow Walls (2011), Of course, I had studied Emerson in literature class, but Weaver suggested that I examine Emerson's work through a scientific lens. Emerson wrote of problems with empirical science:

> Empirical science is apt to cloud the sight, and by the very knowledge of functions and processes to bereave the student of the manly contemplation of the whole. The savant becomes unpoetic. But the best read naturalist who lends an entire and devout attention to truth, will see that there remains much to learn of his relation to the world, and that it is not to be learned by any addition or subtraction or other comparisons of known quantities, but is arrived at by untaught sallies of the spirit, by a continual self-recovery, and by entire humility. He will perceive that there are far more excellent qualities in the student than preciseness and infallibility; that a guess is often more fruitful than an indisputable affirmation, and that a dream may let us deeper into the secret of nature than a hundred concerted experiments. (Emerson, 2003, pp. 74–75)

Quantitative data has a place, but Emerson acknowledges here that not everything can be measured. Imagine how much knowledge can be obtained if we place ourselves somewhere between science and the world.

According to Alexander von Humboldt (1858/1997), science is a journey that promises to "lead us through the vast range of creation that may be compared to a journey in a distant land" (p. 55). Humboldt (1834/1993) himself embarked upon many journeys because he desired to see the world "with [his] own eyes a grand, wild nature rich in every conceivable natural product, and the prospect of collecting facts that might contribute to the progress of science" (p. 15). This collection of facts would not "be regarded as a mere encyclopedic aggregation of the most important and general results that have been collected together from special branches of knowledge" (Humboldt, 1858/1997, p. 55). Such a collection is "nothing more than the materials for a vast edifice and their combination cannot constitute the physical history of the world, whose exalted part it is to show the simultaneous action and the connecting links of the forces which pervade the universe" (p. 55). Collecting, contemplating, and connecting were all part of Humboldt's approach to science. His goals and desires for his practice of science were complex. He tells us that, "each step that we make in the more intimate knowledge of nature leads us to the entrance of new labyrinths" (p. 41). It will be in such mazes, he argues, that "excitement produced by a presentiment of discovery, the vague intuition of the mysteries to be unfolded, and the multiplicity of the paths before us, all tend to stimulate the exercise of thought in all stages of knowledge" (p. 41).

Why so many connections? Simple collections of facts could be detrimental to the progress of science. Humboldt (1858/1997) tells us that the "accumulation of unconnected observations of details … may doubtlessly have tended to create and foster the deeply rooted prejudice, that the study of the exact sciences must necessarily chill the feelings, and diminish the nobler enjoyments attendant upon a contemplation of nature" (p. 40). Don't think for one moment that Humboldt is advocating for soft or less than scientific pursuit of knowledge. His goal was to propagate "an earnest and sound knowledge of science" (p. 53). He believed that "man [sic] cannot act upon nature or appropriate her forces to his own use without comprehending their full extent and having an intimate acquaintance with the laws of the physical world" (p 53). Along with this intense knowledge of nature, Humboldt believed the pursuit of science should be "an inward one—an ennoblement of the intellect—a science that includes "philosophy, poetry and the fine arts" (p. 53). The fields of science:

> Mutually enrich each other and by their extension become connected together in more numerous and more intimate relations, the development of general truths may be given with conciseness devoid of superficiality. On being first examined, all phenomena appear to be isolated, and it is only by the result of a multiplicity of observations, combined by reason, that we can trace the mutual relations existing between them. If however, in the present age, which is so strongly characterized by a brilliant course of scientific discoveries, we perceive a want of connection in the phenomena of certain sciences, we may anticipate the revelation of new facts, whose importance will probably be commensurate with the attention directed to these branches of study. (p. 48–49)

Humboldt argues that it is through the forging of connections that new knowledge is made, and new understandings achieved.

My introduction to the work of Alexander von Humboldt came through the work of Laura Dassow Walls. Walls (2011) opens her book, *Journey to the Cosmos with Alexander von Humboldt*, with a drawing of a natural bridge between two high cliffs in Icononzo, South America. Humboldt (1814/2011) drew this picture himself, sketching "from the northern part of the valley, with a side view of the arch" (p. 60). The natural bridge rests upon two high cliffs, "which upon measuring [Humboldt] found to be, one fifteen hundred, and the other thirteen hundred meters in perpendicular depth" from the valley below (p. 54). Walls (2011) wrote that the bridge's metaphorical construction demonstrated that Humboldt not only wanted to explore the world from a science viewpoint, he wanted to create bridges between "peoples, disciplines, places, and historical eras" (p. 10). Connections were a vital part of Humboldt's method of carrying out science. Can science in classrooms or in educational research become a space, a Humboldtian bridge of sorts, where needed discourses and lingering can take place and connections explored? Could this space be used to demonstrate to students that they are indeed connected to science and science is connected to them?

Emerson (2003) writes that success in science "depends upon elevating science to a creative art—not just marrying science and poetry, but merging the two in a new, prophetic power" (p. 50). To Emerson, "Poetry and science were as close as light and object, speech and air wings and the wind that lifted them" (Walls, 2003, p. 226). In his world, "the only way to recover from corrupting pathologies of social convention was to grow beyond them, as a self and as a community or society, free and self-directed, using the resources of native language, art, poetry and scientific knowledge" combined (p. 235). Is it possible to create such an interdisciplinary relationship? Barone and Eisner (2012) tell us that "Within the 20th century, and now the 21st, regard for this dualism—and, indeed, a science-over-art hierarchy—remains but has increasingly been eroded" (p. x). This erosion has opened the door for new opportunities in research.

When considering the other, science, and art, multiple spaces are created. It will be in these in-between spaces where we can truly discover how science is connected to Henrietta Lacks and how we as teachers and students are able to connect to her story to more deeply understand science itself.

NOTES

1. A fictional academic meeting being held at Johns Hopkins University.
2. A photo of this art display is included with permission in Senior (2011, p. 524). The original art show where this work was presented in 2006.
3. STEAM, stands for science, technology, engineering, art and math education.

CHAPTER 3

ENTER HENRIETTA LACKS

PRESCRIPT: NO STRANGER TO JOHNS HOPKINS

Henrietta Lacks was no stranger to Johns Hopkins Hospital/Research Center. Henrietta Lacks underwent a biopsy there on February 1, 1951. Henrietta's physician, Howard Jones, read her chart before her examination. The chart described her as a "housewife and mother of five" with only a "sixth or seventh grade education" (Skloot, 2010, p. 16). She experienced breathing difficulties and anxiety concerning her "oldest daughter who is epileptic and can't talk" (p. 16). Henrietta reported to the doctor that her household was happy; she was an occasional drinker and not an extensive traveler. The doctor found her to be "well-nourished and cooperative" (p. 16). The chart noted issues with bleeding during her last two pregnancies, "increased cellular activity in the cervix" and a positive gonorrhea test (p. 16). With each of these problems, Henrietta declined treatment. After the biopsy, Howard Jones sent Henrietta home and dictated notes about her visit and her diagnosis. Jones found it "interesting that she had a term delivery here at this hospital, September 19, 1950. No note is made in the history at that time, or at the 6 weeks return visit that there is any abnormality of the cervix" (p. 17). Before her biopsy procedure, the chief resident allowed George Gey to obtain a sample before radium was put in place to treat the very "shiny and purple (like grape Jell-O)" cancerous tumor (p. 17). Jones (1971) reported that the "carcinoma of the cervix was the first to be grown in culture and resistant to radium treatment" (Jones et al., 1971, p. 947). Henrietta died at Johns Hopkins 8 months later on October 4, 1951.

Mandy Hoffen and a Conspiracy to Resurrect Life and Social Justice in Science Curriculum With Henrietta Lacks: A Play, pp. 49–72
49

George Gey died of pancreatic cancer in 1970. During a tribute to Gey, the source of HeLa cells is finally revealed to the public in an article written by Howard Jones and his colleagues (Jones et al., 1971). Jones was the attending physician who diagnosed and prescribed Henrietta's treatment at Johns Hopkins. These colleagues ask a very important question: "Will she live forever if nurtured by the hands of future workers?" (Jones et al., 1971, p. 947). Howard Jones noted that, "even now Henrietta Lacks, first as Henrietta, then as HeLa, has a combined life of 51 years" (p. 947). Using his method of calculation, today HeLa cells have a combined life of 100 years. Those amazing cells have made quite a name for themselves, but Henrietta and her family were not connected to the cells until the Jones et al. (1971) article. Jones combines the age of Henrietta with the age of HeLa cells when it appears scientists doing research with HeLa cells kept the two separate. As a medical doctor who had met Henrietta, Jones may have felt more of a connection to Henrietta. Imagine being a research scientist and your vials of HeLa cells arrive at the lab. Would you feel a connection to the source of the cells? My biology students view human cheek cells, and to their amazement, seeing cells that make up their body helps them feel connected to information taught about cells.

Leigh Van Valen, an evolutionary biologist at the University of Chicago, proposed that HeLa cells in laboratories today are evolving separately from humans, and having a separate evolution is really what a species is all about (Van Valen, 1991). Van Valen (1991) and her group propose that because HeLa cells have been in laboratories and not part of a complete organism, they have acquired different characteristics over time. She proposes classifying them as a new species of cells for several reasons, including that it's "way of life has no resemblance to that of, say, mammals" (Van Valen, 1991, p. 73). Robert Stevenson spent most of his career working with HeLa cells and the issue of contamination in laboratories. His view of HeLa cells and their genetic makeup is different from Van Valen's. In an interview, he told Rebecca Skloot (2010):

> Scientists don't like to think of HeLa cells as being little bits of Henrietta because it's much easier to do science when you disassociate your materials from the people they come from. But if you could get a sample from Henrietta's body today and do DNA fingerprinting on it, her DNA would match the DNA in HeLa cells. (p. 216)

I do not believe that we can think about HeLa cells separate from Henrietta's cells. The source of the cells must be acknowledged. With the Human Genome project, DNA is becoming exactly who we are. At some point in time, we may all carry around our personal genome on a computer chip or SIM card. Our genome is not just a unique sequence of nitrogen bases; our genome is who we are.

In the spring of 2013, the "European Molecular Biology Laboratory in Heidelberg, Germany" posts a portion of the HeLa cell genome on the internet. Here we go again ... without the Lackses' consent (Callaway, 2013). "Imagine if someone secretly sent samples of your DNA to one of many companies that promise to tell you what your genes say about you. That report would list the good news (you'll probably live to be 100) and the not-so-good news (you'll most likely develop Alzheimer's, bipolar disorder and maybe alcoholism)" (Skloot, 2013, p. 2). Why does this lack of consent matter? Let's take it one step further: Once your genome has been analyzed, the results are added to an international database for all to see. Instantly a new venue for discrimination has been invented. "Genetic information can be stigmatizing, and while it is illegal for employers or health insurance providers to discriminate based on that information, this is not true for life insurance, disability coverage or long-term care" (p. 2). Henrietta's HeLa cell genome was posted on the internet without consent. Rebecca Skloot spoke to the great-granddaughter of Henrietta Lacks, Jeri Lacks-Whye. Lacks-Whye stated, "That is private family information, it shouldn't have been published without our consent" (p. 2). Following the release, a scientist "uploaded HeLa's genome to the public website called SNPedia, a Wikipedia-like site for translating genetic information. Minutes later, it produced a report full of personal information about Henrietta Lacks and her family" (p. 2). The scientist only shared the information with Rebecca Skloot. The European team that posted the gene sequence took it down after hearing from the Lacks family. Only 15 people had downloaded the information. The team has hopes of talking "with the Lacks family to determine how to handle the HeLa genome while working toward creating international standards for handling these issues" (p. 4). "The publication of the HeLa genome without consent isn't an example of a few researchers making a mistake. The whole system allowed it. Everyone involved followed standard practices. They presented their research at conferences and in a peer-reviewed journal. No one raised questions about consent" (p. 2).

"The Lackses' experiences over the last 60 years foretold nearly every major ethical issue raised by research on human tissues and genetic material" (p. 2). Think about these issues for a moment. The first and most obvious is that Henrietta Lacks could only get treatment at Johns Hopkins Hospital in a public ward. Secondly, it was common practice among doctors to use patients in the public ward as research material "usually without their knowledge" (Skloot, 2010, p. 29). It was 1957 before the words "informed consent" would make it to a courtroom (p. 132). "Now they're raising a new round of ethical questions for science: though their consent is not (yet) required for publishing private genetic information from HeLa, should it be? Should we require consent before anyone's genome is sequenced and published?" (Skloot, 2013, p. 3).

The story of Henrietta Lacks and her family began over 60 years ago. Through understanding the story and viewing the story through a series of different lenses, the story might prove to be valuable in solving current issues concerning scientific and medical ethics. When will Henrietta become a person deserving of basic human rights? Can Henrietta's story help shape the course of science? What if we all become judged or categorized based on the sequences of our nitrogen bases? What if you are treated as if you are bipolar, because someone sees that possibility reported on some website or database listing your personal genome? Scientists and writers telling the story of HeLa cells often personified characteristics of HeLa cells onto Henrietta herself. The language itself brings additional issues of racism and sexism into the story. The commodification of HeLa cells further complicates the story. Where does ownership stop and start? I can't help but think it is important to contemplate the connections between Henrietta and her HeLa cells for the sake of her family, and for the sake of science.

ACT III

Act III Scenes I and II will involve several characters traveling to The World Alliance of Science and Science Education Symposium at Johns Hopkins Hospital/Research Center and the opening session of this conference. The overarching theme of this imaginary meeting involves celebrating the anniversary of HeLa cells. These first acts present stories of how conference participants journey to Johns Hopkins Hospital/Research Center. As they make their journey, they experience mysterious encounters that are all connected to the story of Henrietta Lacks.

Act III Scene I: Traveling to World Alliance of Science and Science Education

In Act III Scene III Roland Pattillo will be speaking to kick off the meeting. Dr. Pattillo is a professor from the Morehouse School of Medicine. Dr. Pattillo worked in George Gey's laboratory where HeLa cells were originally grown. He was one of the first scientists who set out to tell Henrietta's story. Priscilla Wald, Hannah Landecker, and Rebecca Skloot make up the panel presenting the keynote address. All these individuals have been passionate about the story of Henrietta Lacks. The foundation for other conversations that take place in the following two chapters will be provided in the keynote address. The conversations will allow views to be shared and compared. A resurrected Henrietta Lacks will try to be part of their conversations. Not all members of the panel choose to acknowledge the voice of Henrietta as she tries to share her story and convince the panelists that she and her family should be part of science.

Characters in order of appearance: John Weaver*, professor; Cootie Lacks, Henrietta's cousin; Peter Appelbaum*, professor; and Nancy Cartwright*, physicist.

Travelers make their way to the World Alliance of Science and Science Education Symposium. Today is February 7, 2016. John Weaver finds himself out in the middle of nowhere. He is speaking aloud to himself.

John Weaver: I always get lost when I get in too big of a hurry. The flat tire was a bonus. This place is way past the middle of nowhere. That GPS wench has let me down for the last time …

An old gentleman taking a leisurely walk approaches Weaver. He looks at him closely.

Cootie Lacks*:* Are you one of those Johns Hopkins doctors … here to ask about Henrietta's cells?

Weaver: I am a doctor, but not that kind. I am John Weaver. I am a professor. I seem to have taken a wrong turn off the interstate. I have a flat tire … no cell phone service. This rental car has no spare.

Cootie Lacks: We do not get many of your kind in these parts. (*The old man chuckled*) … No one ends up here unless Hennie makes it happen. Evidently, we are supposed to be having this conversation right now. This is one for the record book: Hennie haunts professor's GPS … Mm mmm. You know, Hennie stopped a reporter from coming to talk to us. He had a wreck. Then his house caught on fire and burned up all the notes from his interview. I got a cousin who can come help you with your tire. My house is right over there … let's say we adjourn to the porch while we wait. I can tell you her story. That is the only reason people end up in these parts. She is buried right down the road. Real nice guy from Atlanta bought her and her little girl Elsie their own headstones, years after they died.

Weaver: When you say Hennie, do you mean Henrietta Lacks? (*Weaver had shared this book with his students. One of his students was very interested in using Rebecca Skloot's book,* The Immortal Life of Henrietta Lacks, *in her classroom. He was in shock … His* *bookshelf was coming to life in the middle of nowhere. Weaver, although agitated, realized he had stumbled into an amazing opportunity. He knew that his best bet was to relax ... be in this moment ... listening to this story firsthand. Maybe Henrietta did want him to be here. It was only a flat tire. He had read about Henrietta controlling life from the grave. He may have been the only traveler to cross Cootie's path in many, many years. The old gentleman was talking nonstop.)*

Cootie Lacks: Of course, but we all called her Hennie. The railroad tracks back there mark the division between the black Lacks families and the white Lacks families. Henrietta's grandfather was white. Her mother was a slave. They farmed these parts together, and then eventually became landowners.

Weaver: What can you tell me about Johns Hopkins and those immortal cells?

Cootie Lacks: I will get there ... You have to listen ... unless you are going to walk ... it kinda looks like to me that you have no better place to go.

Weaver: I cannot argue with that. Lead the way. I am listening! You got any cold Mountain Dew ... the kind with real sugar? (*Cootie threw his head back and laughed.*)

Cootie Lacks: Don't you worry any. I will take good care of my new doctor friend. Henrietta would not have it any other way.

Evening of February 7, 2016—welcome to World Alliance of Science and Science Education Symposium reception being held in the hospital lobby.

Peter Appelbaum: Hi John, I am glad you made it. I got a short distress text from you ... the only word on the screen was "flat" ... I figured if you needed me, you would have sent more info.

Weaver: Honestly, Peter, I expected more of you. Maybe a squadron of search helicopters, a full-scale ground operation.

Appelbaum: (*Chuckles.*) You remember David Blades and Nancy Cartwright. I am glad that you have already had an epic adventure just getting to the conference. Maybe after the reception we can all grab a drink and you can tell us all about it. We have not seen Patti yet. I know she is coming. Maybe we should send out a search party for Patti.

Weaver: I met the most interesting man. He thought I was a doctor from Johns Hopkins Hospital coming to ask questions about Henrietta Lacks.

Appelbaum: Henrietta? Isn't she the woman those cells came from—the reason the meeting is at Johns Hopkins this year? She was a patient here at Hopkins, right?

Weaver: This guy tried to convince me that Henrietta had risen from her grave to make sure I had ended up in Pleasantville. He wove a great story and had a refrigerator of cool drinks. I rate it as one of the best detours I have ever taken.

David Blades: I did not know about Henrietta until your graduate student sent me an email about her work. (*He stopped abruptly and snapped his fingers.*) I met her today. It was dark ... we were all standing around and talking about a HeLa exhibit. She was talking about Henrietta Lacks. That had to be her. She had some questions about my STEM play. The group had a very interesting discussion going on around a BioArt exhibit.

Nancy Cartwright: You are part of her dissertation committee, right, John? She wanted to know if I thought Henrietta Lacks should be part of science curriculum. John, it is priceless that you ended up in Henrietta's town talking to Henrietta's family, while serving on a dissertation committee for a student writing about Henrietta Lacks. (*Nancy begins to hum the Twilight Zone theme.*) It is experiences such as your detour that make our "dappled world, a world rich in different things, with different natures, behaving in different ways" (Cartwright, 1999, p. 1). Better watch out. Henrietta might have a hold on you ... from the grave. (*The group laughs.*)

Weaver: To a dappled world!

The four raise their glasses in a toast.

Act III Scene II: Meeting Day

Characters in order of appearance: Bruno Latour*, philosopher of science; Michel Serres*, philosopher of science; Donna Haraway*, philosopher of science; young child about 5 years old; young child about 3 years old; Father; Patti Lather*, university professor; Henrietta Lacks.

A group of professors are making their way to a meeting at Johns Hopkins University. They all have an encounter with the story of Henrietta Lacks at some point in their journey.

Today, February 8, 2016, was no different than any for Bruno Latour. It did not matter where he was; his routine was always the same. He spent the first part of each day carefully examining his morning papers. Today he carefully analyzes three papers: *The Baltimore Sun, The USA Today*— kindly provided by the hotel staff, and *The New York Times*. Bruno Latour gathers his newspapers and examines their contents in a way that will make a concrete sequential reader's head spin. Latour jumps from an idea on one page to an idea on a previous page in a back and forth fashion to show how the headlines in no way reveal a cohesive story. During this morning's exploration, Bruno first reads an article about a lawsuit being settled in Massachusetts, granting a family the right to deny researchers access to their mother's cells. On page 3, he reads the most recent best seller list and notes that *The Immortal Life of Henrietta Lacks* has remained on this list for over 3 years. He sees that a scientist, Hans Jörg Rheinberger, is currently in town to present a paper on how RNA translation takes place. The next headline reads: "Experimental science not all it is cracked up to be?" Hmm, thought Latour to himself. It is a low blow to current close-minded educational policymakers ... school science is monopolized by learning a simple experimental method. Unbeknownst to him, the connection to the best seller and the ownership of cells will become part of his day. He heads downstairs to the hotel restaurant for one more cup of coffee with his longtime friend and colleague Michel Serres.

It is very unusual for Michel Serres to agree to any public event. He decides to make this trip due to the encouragement of Bruno Latour. He was on faculty at Johns Hopkins years ago. He and Latour share a common interest in science. The pair collaborated on a book several years prior to this event. His session will allow him to discuss the importance of philosophy, history, culture in science, and connections to humanities and literature. This is a rare opportunity at a research institution such as Johns Hopkins, where traditions are strong in strict scientific methods. Serres' life was formed by science, a science unrecognizable in today's world of high-gloss, high-priced science ventures.

The same morning, coming from the west end of town, Sandra Harding enters the Johns Hopkins Hospital lobby. The lobby is serving as an exhibition hall for publishers and other displays filled with glitzy projects like personal genetics and personal medicine. The large company Moneysanto is giving away computer tablets as well as food samples, composed of genetically modified organisms they produce, to all educators in attendance. A large group of pharmaceutical companies fills the halls with promises that their medications improve test scores. There is even a booth advertising teacher robots. Now I have seen it all, Sandra thinks to herself. Over in the corner, away from the fanfare, she sees a tattered display cabinet, obviously part of the actual lobby, with pictures of a basement laboratory. The caption reads: "It was here that George Gey and his assistant Mary Kubicek were able to grow the first human cells outside the human body." There were pictures of HeLa cells being packaged and sent all over the world. A small model of George Gey's drum roller was in the case. Seeking solitude away from the center ring of this circus, Harding begins to read the text posted on the back of the display cabinet. George Gey, a man who never patented his famous drum roller or sold HeLa cells, was at the center of a huge controversy involving the taking of tumor cells following a woman's surgery. The woman was a wife and mother of five, just 30 years old. The controversy included a huge cover-up. Although she is intrigued, Sandra glances at her watch and decides she'd better hurry to her meeting room.

Donna Haraway is making her way to Johns Hopkins Hospital through the Baltimore morning rush hour traffic. Science teachers and leaders in science education were coming from all over the world to attend this conference. Donna, a university professor, had arrived the evening before. The hotel shuttle dropped her off across the street from the hospital entrance. The driver gave her directions to the conference building. She decided to sit on the bench under the giant oak and enjoy her coffee before starting her day. As she closed her eyes, she sighed ... she had slept restlessly in the hotel the previous night. Honestly, she had expectations of this being just another long-winded session of mostly men, disconnected from humanity, women, science, and classrooms, giving lofty advice. Sitting on the bench with her eyes closed, just resting for a moment, she hears a small child's voice:

Child: Daddy, when we goin up to see Mama?

Father: Child, she is right there in the window ... can you wave at her? She is too sick to come down and too sick for you to go up. This is as close as we can go ...

Haraway opens her eyes, expecting to see a child and her father, but there is no one. Maybe they had walked quietly past. She stands up and looks at the wall of windows. She cannot see anyone waving. She chuckles ... It was probably a dream ... me all the way in Baltimore ... no worries, nothing to do for a whole 30 minutes ... I probably had myself a nap and did not know it. She decides to sit back down at this lovely place. Eyes closed again ... she refuses to be nervous about the upcoming meeting.

Child: Daddy, make little Deborah stop crying … maybe you need to change her diaper or hold and rock her like mama does."

Father: I will pick her up, Lawrence. Can both of you look way up there? See Mama, she is looking out that window to see you.

This time Haraway knows she was not asleep … she jumps up and whirls around … no one is there. The leaves–dull and brown, lacking their vibrant fall colors— are rattling as a cool winter breeze filled the air. The breeze gives her a chill. Okay, I need to move on, she decides. She crosses the street and entered the hospital lobby, looking for a very strong cup of coffee.

Patti Lather's cab arrives outside of Johns Hopkins Hospital at 8:30 A.M. Her meeting was on the fourth floor at 9:00 A.M. I can make it, she thinks. Her flight had been delayed overnight, and there were no polite words for the bustling Baltimore traffic. She hurries through the entrance of the Johns Hopkins Hospital lobby. A historical display of photographs catches her eye. In the pictures, she notes the colored rest room and a large sign pointing to the colored ward. Then she notices a group of old photographs of a beautiful African American woman … hands on hips … bright red lipstick. The caption under the picture revealed that her name was Henrietta Lacks. Completely distracted from the original goal of making her 9:00 meeting on time, Patti stops, pulls her tablet from her bag, and Googles the name: Henrietta Lacks. She clicks on the first article and begins reading: Her name was Henrietta Lacks but the world knew her as HeLa. Patti had studied HeLa cells in a college biology class … but she had never even questioned their origin. Why not? Now it seems like a missed opportunity to have not asked more questions in her science classes. The second link is about a book … she sits down in the lobby and begins to read an excerpt. She looks up from the device and sees a beautiful African American woman come through the revolving door wearing a stylish brown suit. A bone-chilling breeze blew through the lobby. The woman stops at the information desk.

Henrietta Lacks: Can I get directions to the public ward?

Receptionist: (*She did not hear Henrietta's question. Everyone had been asking the same question, so she repeated her answer*): The presentation hall. Go straight and take the elevator on the right up to the fourth floor.

Another cold chill runs through Patti's body as she looks up from her device again and toward the information desk. This was 2016—there are no public wards in hospitals. She blinks and decides she'd better make her way upstairs. There was not time to question the receptionist … but she would be back later.

Act III Scene III: Opening Session

Characters in order of appearance: Roland Pattillo*, professor of medicine; Patricia Wald*, professor of literature; Rebecca Skloot*, author; Hannah Landecker*, professor of biology; Henrietta Lacks; Patti Lather*, professor of Curriculum Studies; John Weaver*, professor of curriculum studies; Donna Haraway*, philosopher of science.

By 9:00, all four scholars are sitting quietly in the auditorium, waiting on the keynote address to begin. There is not much small talk in the auditorium. Everyone seems very perplexed ... even quiet ... considering how scholars usually behave at academic meetings. As others file into the auditorium, three women walk quietly up the stairs to the stage. The women take their places in chairs on the right side of the podium. Each adjusts her microphone; each situates her water glass. Oddly enough all three are wearing an article of red clothing. Roland Pattillo makes his way to the podium from the other side of the stage.

Roland Pattillo: Good morning. It is my honor to be here to welcome you to this year's meeting of The World Alliance of Science and Science Education. This year we are not only gathering for discussion of science and science education, we come to honor the anniversary of the HeLa cell line. In recognizing the importance of this cell line, we also wish to pay homage to the woman, Henrietta Lacks, who served as the source or donor of these cells. This tremendous story will open many opportunities for discourse. The theme of this year's meeting is very significant to me personally. As a young "postdoctoral fellow in [George] Gey's lab," I found myself immediately connected to Henrietta Lacks (Skloot, 2010, p. 219). Science can tell the story of the cell and the cell lines, but I found myself compelled to tell her personal story.

You will not be hearing a lot of statistics and empirical data concerning HeLa cells today. We are here to celebrate the woman behind the cells—the stories of the woman and the cells. We hope that when you continue your discussions in the breakout sessions later, you will find the conversation centering on Henrietta and her HeLa cells, not simply the latter.

Our keynote speakers will be sharing a tremendous story with you today. A story that includes racism, economics, and what it means to be a human amidst all the new advances in biotechnology. It is my honor to introduce them. First on my left, we have Priscilla Wald. Dr. Wald currently teaches at Duke University. She "teaches and works on U.S. literature and culture, particularly literature of the late-18th to mid-20th centuries, contemporary narratives of science and medicine, science fiction literature and film, and environmental studies" (Duke University, 2015). Wald's "work focuses on the intersections among the law, literature, science, and medicine" (Duke University, 2015). She has written about Henrietta and Henrietta's HeLa cells. Her current interests involve tackling issues brought about by the Human Genome project.

Next, we have Hannah Landecker. Dr. Landecker is currently the acting director of the Institute for Society and Genetics at UCLA. She "uses the tools of history and social science to study contemporary developments in the life sciences, and their historical taproots in the twentieth century" (UCLA Institution for Society and Genetics, 2015). Hannah "has taught and researched in the fields of history of science, anthropology and sociology [focusing] on the social and historical study of biotechnology and life science, from 1900 to now" (UCLA Institution for Society and Genetics, 2015). She authored a book called *Culturing Life: How Cells Became Technologies* (2007). For those of you interested in a detailed historical account of how HeLa cells became the first human cell line, I recommend it highly.

Finally, we have Rebecca Skloot. In academia, we would all be lying if we said we have not all wanted to accomplish what this science journalist has accomplished. She has brought the story of Henrietta Lacks to the masses. A story that forever changes anyone that hears it. Change does not happen because the story of Henrietta Lacks is so unique. Unfortunately, there are many similar stories; change is possible because the story is written in a way in which we can connect to the woman behind the cells. What will change when you meet Henrietta and her family? Will it be one's willingness to connect to the world around them? If you have not heard of Rebecca's book, it might be because you stay holed up in your laboratories … working hard…isolated from the world around you. We as researchers must do diligent work and keep long hours, but we must not neglect to keep up with the needs of the changing world around us. The *Immortal Life of Henrietta Lacks* began selling in January of 2010. The novel has spent over 2 years on the *New York Times* Best Seller List (Best Sellers, 2013). Skloot spent 10 years unearthing the story of Henrietta Lacks, the woman behind the amazing HeLa cells, and the first human tissue culture line for medical research. (*Pattillo takes out a copy of* The Immortal Life of Henrietta Lacks *from under the podium.*) Ms. Skloot opens her book with a quote from Elie Wiesel, writer of *The Nazi Doctors and the Nuremberg Code* that I would like to share with you. I think it sums up our collective purpose here today very well: "We must not see any person as an abstraction. Instead, we must see in every person a universe with its own secrets, with its own treasures, with its own sources of anguish, and with some measure of triumph" (Wiesel, 1992, p. ix). Skloot's passion was to unearth Henrietta's story and put a name and face with the story of HeLa cells. All three women who share the stage with me today share the mission of telling the story of Henrietta Lacks. Ladies, I turn the microphones over to you.

Priscilla Wald: I like the quote, Rebecca—what a great quote to open your book and get to the heart of the matter. Dr. Pattillo, I think that is a great way to focus our conversations today. No one should be an abstraction. "This is a global story, and is also a biotechnological story, which is to say that it entails the development of laboratory techniques and associated business interests that involve the production, use, and marketing of living organisms" (Wald, 2012a, p. 185). It is an honor to be with all of you here today.

Skloot: I want to thank you personally, Dr. Pattillo, "for taking time to figure me out, for believing in me, for schooling me and for helping me contact Deborah Lacks" (Skloot, 2010, p. 339). You and your wife opened your home and even served an editorial role for *The Immortal Life of Henrietta Lacks. (She looks across to the other members of the panel.)* I agree it is an honor to be here today, under the same roof where it all happened. The three of us got together yesterday when we all arrived in Baltimore to discuss our presentation. We did not want to duplicate each other's efforts. We each hope to tell a different portion of the story. I will apologize for any overlaps up front. We each bring a different perspective to this story, but we share a common purpose: This story needs to be shared, not only to honor Henrietta Lacks and her family, but the application of this story to current-day issues in science and science education might help us identify specific areas that need resolution. We thought it would be helpful if you heard the history of the HeLa cell line first. Then we can get into the issues that came about from the creation of the cell line. We are not going to be very formal about our speaking. We will all interrupt and interject when we feel necessary. Hannah, why don't you get this story started.

Hannah Landecker: Good afternoon. For those who have been holed up in their laboratories and are unfamiliar with this story, I am going to offer a very brief summary. George Gey began his work trying to grow cells outside the human body, right here at Johns Hopkins Hospital/Research Center some 60 years ago, down in the basement. He was specifically committed to the project of growing "cervical cancer cells in the hope that their life under glass would reveal something about their action in the body" (Landecker, 2000, p. 57). Gey was looking for a cure for cervical cancer. He felt that if he could study cells outside the body, he could find the answer. In February of 1951, "A piece of cancerous cervical tissue was cut from a woman named Henrietta Lacks" (Landecker, 2000, p. 53). Lacks died on October 4, 1951. "Without her knowledge or permission, Lacks became part of the cervical cancer research project when a piece of the biopsy material was sent to the Gey laboratory" (p. 53). Mary Kubicek, George Gey's laboratory technician, had been working for 2 years to grow cells in culture. Each sample of "live cells [from the biopsies] were grown in test tubes, supplied with nutrient medium and kept at body temperature in an incubator" (p. 53). Although for years the donor's identity was kept a secret ... HeLa was thought to stand for possibly Helen Lane, or Helen Larson. "Researchers realized that this cell line would do what others would not—continue to grow and divide, quickly and copiously" (p. 128). "George Gey never attempted to patent, or otherwise limit the distribution of HeLa cells, clearly not anticipating the chain letter effect of sending out cultures which were grown up, split into parts, and sent on to others" (p. 57). The story has two endings: "the cells live, and the woman does not" (p. 53).

As Henrietta's family is moving on and living a life without her and without knowledge of her immortal cells—life is good in the tissue culture world ... for the moment. "HeLa cells were being mass-produced as part of a push for the rapid evaluation of the polio vaccine, which Jonas Salk developed in 1952" (Landecker, 2000, p. 57). "The Tuskegee Institute, a historically black college in Alabama, was appointed by the National Foundation for Infantile Paralysis to be the locus of production" of HeLa cells (pp. 57–58). The use of Henrietta's cells became widespread, and the line "began to stand in for a generalized human or cellular subject...a factor in this flourishing of research was the availability of living human materials for experiments that could or should not be performed on living persons" (Landecker, 2007, p. 165). "The ubiquity of HeLa continued, but its invisibility as 'standard reference cell' faltered in 1967 with the announcement that HeLa cells had contaminated and overgrown many of the other immortal human cell cultures established in the 1950s and 1960s" (p. 168).

Skloot: I would not have wanted to be Stanley Gartler on that day at the "Second Decennial Review Conference on Cell Tissue and Organ Culture" (Skloot, 2010, p. 152); I imagine the setting would be similar to what you are experiencing today. An audience filled with scholars and cell biologists. "There in front of George Gey and the other giants of cell culture, Gartler announced that he'd found a 'technical problem' in their field" (p. 152). Gartler informed the audience "all those years researchers thought they were creating a library of human tissues; they'd probably just been growing and regrowing HeLa" (p. 154).

Landecker: Gartler tested 18 different cell lines for a group of enzymes that differ very slightly between people. "All 18 had the same profile as the HeLa cell. The key piece of evidence in this study was the profile for a particular enzyme called G6PD (glucose-6-phosphate dehydrogenase), which is a factor in red blood cell metabolism" (Landecker, 2007, p. 168). Gartler (1967) told the audience, "I have not been able to ascertain the supposed racial origin of all 18 lines; it is known, however, that at least some of these are from Caucasians, and that at least one, HeLa, is from a Negro" (p. 173).

Wald: It is the accounts of this technical problem that bring to light the complicated issues surrounding the HeLa cell line. Gartler contacted George Gey and explained the problem. He believed that cross contamination had occurred in laboratories where new cell lines were being established. Some of the cell lines had a specific protein marker. "Racial identification of the donor of the HeLa cells became important when Gartler sought to identify the extent of the contamination of other cell lines" (Wald, 2012b, p. 250). Remember, HeLa cells were the first cells to grow outside the body. Now it is "the very properties that had allowed the cells to survive in the petri dish in 1951 [that] also posed a serious problem for medical researchers. The remarkable aggressiveness of the HeLa cells made them very difficult to contain, and they contaminated other cell lines, in some cases invalidating years of research (p. 250)." Gartler had to know the race of the donor. Why, you may ask?

> The donor's racial identification was relevant to Gartler for tracking the cells: it had no bearing on their aggressiveness. The anthropomorphism of the cells shows both how the race and gender of the donor permeated the public narrative and how cultural biases surface when scientific research yields new entities. (Wald, 2012b, p. 250)

Landecker: We must begin wondering where Henrietta begins and ends in reference to HeLa cells. When listening to descriptions of HeLa cells, "HeLa cells have been continually personified in the image of their origin, Henrietta Lacks" but these personifications change as time passes and the story evolves (Landecker, 1999, p. 212). In the beginning when HeLa cells were depicted "as a miracle of modern science and part of the triumph over polio, HeLa was personified with the image of a beneficent young Baltimore housewife" (p. 212). Most of this personification grew out of how science was writing about HeLa cells. The narrative written and published in medical textbooks and professional journals with "Lacks's photograph side by side with photographs of HeLa cells in culture in locations such as medical textbooks and professional journals. These narratives were then taken up with enthusiasm by journalists" (p. 212).

In 1967 when Gartler made his announcement concerning HeLa contamination, we see Henrietta personified in a different way. Gartler, a geneticist, addressed the group of molecular biologists by saying, "HeLa cells originally came from a 'Black' donor, and produced a protein from a gene allele that he said was specific to the 'Black population'" (Landecker, 1999, p. 213). If this protein specific to Henrietta's HeLa cells was found in a cell line from a white person, scientists used the word contamination. The language used continued to digress, "with this announcement, the personification of HeLa changed instantly to that of Henrietta Lacks as a promiscuous, malicious black woman" (p. 213). "Unlike the writers in the 1950's, these authors were not interested in the figure of the self-sacrificing housewife" (Landecker, 2007, p. 171). HeLa cells were now described as having an "identifiable biological race due to their particular genetic structure as black," "female," "vigorous," "aggressive," "surreptitious," "a monster among the Pyrex," "indefatigable," "undeflatable," "renegade," "catastrophic," and "luxuriant" (p. 171). The narrative of reproduction out of control was linked with promiscuity through references to the cells' wild tendencies and their "colorful laboratory life" (p. 171).

Around 1980, we see another shift in descriptions of HeLa cells, "when the perception of an economic value for cell lines refocused attention on the circumstances of the cell line's origin in a biopsy for which informed consent was neither asked nor given" (Landecker, 1999, p. 172). Now the issue of interest dealt with laws of consent and questions of ownership. HeLa cells "should have had a dollar value from the beginning, because look what they would be worth today, after all these years the investment account that is the burgeoning biotechnology industry" (p. 172). HeLa cells are still being sold today. Rebecca, do you know the current dollar amount for a vial of HeLa cells?

Skloot: "Invitrogen sells HeLa products that cost anywhere from $100 to nearly $10,000 per vial" (Skloot, 2010, p. 194). What is most interesting to me in this discussion of profits is the number of individuals who have profited because of HeLa cells. You should all take a minute on your computer and search the "U.S. Patent and Trademark Office database" for patents related to HeLa cells (p.194).

Members of the audience begin to reach for their tablets; Dr. Pattillo on the stage hits send on his search.

Skloot: What did you get, Dr. Pattillo?

Pattillo: "More than 17,000 patents involving HeLa cells" (Skloot, 2010, p. 194).

Wald: Amazing! Prior to 1974, Henrietta's family does not even know that HeLa cells exist. "Many have profited from the HeLa cell line, [but] Lacks's family has not shared in any of the profit" (Wald, 2012b, p. 247). Many familiar with the case, including medical ethicists and Lacks's family, have a perception of wrongdoing on the part of the medical establishment. Throughout the history of this story, the medical establishment here at Johns Hopkins has been questioned concerning wrongdoing. Discussions of the story center on "the racism evident in the treatment of Lacks and her family and the narratives that circulated when the identity of the donor of the HeLa cells became public knowledge. It is, however, difficult to pinpoint the exact nature of the violation" (p. 247).

Skloot: "Spokespeople for Johns Hopkins, including at least one past university president, have issued statements to me and other journalists over the years saying that Hopkins never made a cent off HeLa cells since George Gey gave them away for free" (Skloot, 2010, p. 194).

Wald: The case of Henrietta Lacks did not have much precedence. It was commonplace for scientists to take tissue samples without consent forms in those days. The family struggled, while medical science and pharmaceutical companies were able to profit from the prolific cell line. If the cells had been patented shortly after they began growing in laboratories, history may have been different. "Unlike subsequent researchers who patented cell lines, the Geys did not get rich from the HeLa cell line" (Wald, 2012b, p. 248).

Skloot: I don't think George Gey was ever in the business of growing cells for profit. He did not patent his "roller-tube culturing technique" (Skloot, 2010, p. 39). Without this invention, cells would not have grown in culture.

Wald: The issue of patents and profit are secondary to the issues of racism present in this story. Henrietta died in a segregated ward. Her cells were taken without permission and "used to develop drugs and other treatments to which many of her decedents might not have access" (Wald, 2012b, p. 248). I don't think it makes any sense to "argue that Lacks should not have been on a segregated ward because her cells were unique or that her family should have access to better health care because of the properties of Lacks's cells" (p. 248). What the story of Henrietta Lacks shows us is "the institutional racism that has plagued the nation since its inception. Lacks should not have been on that public ward because that public ward should have never existed" (p. 248). With that said, don't you think that she and her descendants should have access to better health care because such access "ought to be a fundamental human right and not a privilege?" (p. 248).

Skloot: Considering racial issues, economic issues, power, and privilege, I think the last part of this story we must discuss has to be bioethics. Let's go back in time to 1947, when "a U.S.-led tribunal in Nuremberg, Germany, had sentenced seven Nazi doctors to death by hanging. Their crime was conducting unthinkable research on Jews without consent: sewing siblings together to create Siamese twins, dissecting people alive to study organ function" (Skloot, 2010, p. 131). The Nuremberg Code, a code of ethics with 10 specific points, was written to provide guidelines to supervise all human experimentation throughout the world. "The first line in the code says, 'the voluntary consent of the human subject is absolutely essential'" (p. 131). Asking for consent was a revolutionary idea. "The Hippocratic Oath, written in the fourth century BC, did not require patient consent. And though the American Medical Association had issued rules protecting laboratory animals in 1910, no such rules existed for humans until Nuremberg" (p. 131). "The term 'informed consent' first appeared in court documents in 1957 ... decades before anyone thought to ask whether informed consent should apply in cases like Henrietta's where scientists conduct research on tissues no longer attached to a person's body" (p. 132).

Henrietta's family came into this entanglement in 1974. A researcher at Johns Hopkins called Day Lacks, Henrietta's husband, and asked him to gather all the children at his home so they could come and draw blood. The researchers, Victor McKusick and Susan Hsu, were only hoping to test for specific HLA markers that could be used to identify Henrietta's HeLa cells in the laboratory. There was a huge language barrier between Day Lacks and Hsu, the Chinese research fellow. When the researcher called, Lacks thought that they were coming to the house to draw blood to test the children for cancer. He called all the children to meet with the researchers the next day. No attempt was made to obtain informed consent. Deborah, Henrietta's daughter, was "now almost 24, not much younger than Henrietta had been when she died. It made sense they were calling saying it was time for her to get tested" (Skloot, 2010, p. 184). Deborah had been preoccupied with her mother dying of cancer at age 30. She suspected that she would die of cancer when she was 30 as well. She kept calling, asking for the results of the cancer test. She would not stop worrying. She was terrified that she might have cancer and consumed with the idea that researchers had done—and were perhaps still doing—horrible things to her mother. Deborah began studying, asking questions. I was with Deborah the first time she saw her mother's cells. Deborah worked hard to learn about her mother's cells and her mother's cancer. Let me read you an excerpt from Deborah's journal. *(She focuses on her notes)*

Going on with pain

... we should know what's goin on with her cells from all of them that have her cells. You might want to ask why so long with this news, well it's been out for years in and out of video's, papers, books, magazines, radio, TV, all over the world.... I was in shock. Ask, and [no one] answers me. I was brought up to be quiet, no talking, just listen.... I have something to talk about now, Henrietta Lacks what went out of control, how my mother went through all that pain all by herself with those cold-hearted doctors. Oh, how my father said, how they cooked her alive with radiation treatments. What went on in her mind all those months, not getting better and slipping away from her family? You see I am trying to relive that day in my mind. Youngest baby in the hospital with TB, oldest daughter in another hospital, and three others at home, and husband got to, you hear me, got to work through it all to make sure he can feed his babies. And wife dying ... Her in that cold looking ward at [John Hopkins] Hospital, the side for Blacks only, oh yes, I know. When that day came, and my mother died, she was robbed of her cells, and John Hopkins Hospital learned of those cells and kept it to themselves, and gave them to who they wanted and even changed the name to HeLa cell and kept it from us for 20+ years. They say Donated. No No No Robbed Self. My father has not signed any papers ... I want them to show me proof. Where are they? (Skloot, 2010, p. 195)

As Skloot reads, the door in the back of the auditorium squeaks open. An African-American woman in a brown suit slips into the shadows and takes a seat on the back row.

Henrietta Lacks: Did my precious Deborah write that?

Skloot: Excuse me, did someone have a question? *(She looks around, and squints trying to see where the voice came from.)* I think Deborah's writing sums up the issues this story—racism, economics, class, power, and ethics.

Landecker: The connection between Henrietta and her immortal cells is where this story starts to get interesting. Some think of the two as interchangeable. "They somehow stand for her and her for them: otherwise this situation would not present itself as a paradox, much less one that has generated such fascinated attention from 1951 to today" (Landecker, 2000, p. 53). "Cell lines are made to stand in for persons in the first place: They function in the laboratory as proxy theaters for intact living bodies" (p. 54). HeLa cells can function in the following capacities:

> Sites of manufacturing of viruses or proteins or antibodies—cell lines are the tools of the industry whose product is human health. Their identification as "living" and "human" entities cannot fall from them, because it is this origin that gives them commercial and scientific value as producers of biological substances for use by humans and their validity as research sites of human biology. (Landecker, 2000, p. 54)

Skloot: It is easy to view stories as a having a beginning, middle, and successful conclusion. We can be caught up in all the good stuff … without seeing the entire story. Some people never get past this shortened version: That HeLa cells have been "bought, sold, packaged, and shipped by the trillions to laboratories around the world" (Skloot, 2010, p. 2). HeLa cells "went up in the first space missions to see what would happen to human cells in zero gravity" and were crucial in "developing the polio vaccine, chemotherapy, cloning, gene mapping and in vitro fertilization" (Skloot, 2010, p. 2). We hope as you move into the rest of your discussions, that issues of race, economics, and what it means to be human when dealing with cell lines will make it into your conversation. What does Science have to gain from including Henrietta in this narrative? Can Henrietta Lacks herself become the center of this narrative, for science, for education, for humanity?

The keynote session ends, and everyone begins to depart to their assigned session rooms.

Patti Lather: *(turns to John Weaver)* Did you see the woman sitting on the back row?

Weaver: Which one?

Lather: She had on a brown suit, sandals—a very pretty woman. I saw her in the main lobby this morning. She asked where the public ward was. I will be at our session in just a moment. I need to run downstairs first.

Weaver: Are you sure she said public ward? That is odd, there has not been a public ward here for years. Hopkins established the hospital to help the poor. In his will he set aside money specifically to help Black orphans. But I am not sure his ideals were upheld. My new friend Cootie Lacks told me that Hopkins's relationship with the Black community in Baltimore was not good. Cootie reported that the doctors were using children for experiments. He also said that people were afraid to go out at night because doctors from Hopkins might grab them and take them away to their laboratories.

Donna Haraway: Cootie Lacks? (*Haraway gives Weaver a very puzzled look.*)

Weaver: It is a long story; I got a flat tire on the way to Baltimore and met a member of Henrietta's family. Just a chance meeting. He helped me get my tire fixed. We sat on his porch and talked. Mostly about Henrietta. I better get moving, I need to get the room set up for my session. ·

Haraway: I thought my trip was strange. Patti, I saw the woman you are talking about. She looks a lot like Henrietta. Maybe she is a relative.

Lather: I am going to see if the receptionist remembers her.

POSTSCRIPT: ENCOUNTERS WITH HENRIETTA

In the process of creating a conspiracy and seeking avenues for change in education, I have chosen to cross the boundaries of simple narrative. First, I employed conversations, then narrative, storytelling, and playwriting, and science fiction. If I were just telling the story of Henrietta Lacks, or just chronicling my life as an educator, there would be no need ... no purpose. Retelling stories can provide a time of reflection, but the idea is to create a conspiracy to open up new ways of thinking in order to change education. Science fiction "provides a genre, a medium through which the future can be speculatively visualized in the present" (Weaver et al., 2004, p. 1). From an educational point of view, "science fiction can also open up students'/teachers' minds to previously unforeseen possibilities while concurrently empowering people to become curricular creators and cocreators as well as theorists" (p. 1). Science fiction allows us to travel a path in which "we are freed to take flight, soaring away from the present course of essentializing educational practices, proletarianizing policies, commodification, exploitation, and objectification ... while seizing (imaginative) power to influence the way in which education is visualized in the present" (p. 3).

In Act II Henrietta comes to life from a beaker of cells and walks into an imaginary conference. Before entering the conference, Henrietta reclaims her own clothes from a wax mannequin standing in an exhibit hall. Deborah, Henrietta's daughter, resisted the idea of a wax figure on display in a museum when Henrietta's identity was first made known and the first foundation for the Lacks family was proposed: "the family don't need no museum, and they definitely don't need a wax Henrietta," she said. "If anybody collecting money for anything, it should be Henrietta children collecting money for going to the doctor" (p. 223). Henrietta reclaiming her clothes allows her to walk into the meeting as Henrietta, not just HeLa.

In Act III, we see that the scholars have all encountered Henrietta Lacks in some way during their travel to the meeting at Johns Hopkins. Their interest has been sparked by something that took place outside the meeting itself. Each has listened to individuals who are passionate about telling the story of Henrietta Lacks. They are ready to discuss the story of Henrietta along with the story of the amazing HeLa cells. Henrietta being brought back to life allows her to become part of these conversations as a person, not just represented as HeLa. Michel Serres (1991/1997) tells us that

> Science speaks of organs, functions, cells, and molecules, to admit finally it's been a long time since life has been spoken of in laboratories, but it never says flesh, which very precisely, designates the mixture of muscles and blood, skin and hairs, bones, nerves and diverse functions, which thus mixes what the relevant disciplines analyze. (p. xvi)

Henrietta's cells have been a major topic in science writing and science laboratories since George Gey announced their incredible growth on prime-time television. Henrietta's name was hidden. It was all about George Gey's HeLa cells. Serres is telling us that "flesh" should be part of this story even in the laboratory (Serres, 1991, 1997, p. xvi). What part of the story should become part of science? Henrietta Lacks the Baltimore Housewife? Henrietta as HeLa? Or both? Maybe each should be considered, and a new space created in order to examine how Henrietta Lacks and her story should be included in science curriculum. As you begin to hear the story and learn about Henrietta and her cells, I must offer some words of advice. Make sure you are open to the whole story. You may want to forget everything you know about Henrietta Lacks up to this point and start fresh. My fear is that people who hear the story may only be remembering aspects of the story such as HeLa cells went to the moon, or that they can "wrap around the earth at least 3 times, spanning more than 350 million feet" (p. 2). The purpose of sharing this story is not to simply bring recognition, honor and respect to our cell donor, Henrietta Lacks. The purpose of telling the story is to examine and linger within the story and see how it can be applied to the problems we find in current-day science education.

CHAPTER 4

CONSPIRING AND INTERRUPTING PHILOSOPHY AND HISTORY OF SCIENCE

PRESCRIPT: NOT JUST HELA, HENRIETTA LACKS

Rebecca Skloot begins her prologue in *The Immortal Life of Henrietta Lacks* by describing a photo that hangs on her wall of a woman she has never met. "She looks straight into the camera and smiles, hands on hips, dress suit neatly pressed, lips painted deep red ... oblivious to the tumor growing inside her—a tumor that would leave her five children motherless and change the future of medicine" (Skloot, 2010, p. 1).

In the beginning, only researchers who were having issues of contamination with their cell lines needed or wanted to know Henrietta's story. Once Stanley Gartler announced Henrietta's identity and race at a 1966 scientific conference. Michael Rogers (1976), Michael Gold, (1986), Octavia Butler (1987), and Hannah Landecker (1999, 2007) all published "public narratives about Lacks that put assumptions about race, sex, and gender conspicuously on display" (p. 247). In addition to these major publications, The British Broadcast Company made a documentary called "The Way of All Flesh" directed by Adam Curtis (1997). It was not until Rebecca Skloot published *The Immortal Life of Henrietta Lacks* that Henrietta gained attention made possible via a *New York Times* bestseller. Now

Mandy Hoffen and a Conspiracy to Resurrect Life and Social Justice in Science Curriculum With Henrietta Lacks: A Play, pp. 73–103
Copyright © 2021 by Information Age Publishing

that everyone is reading about Henrietta Lacks, one must ask how this text, with such a history, will affect the teaching of science. Can Henrietta's voice be heard in such a conversation?

ACT IV

In Act IV, Henrietta Lacks works to make her way into a conversation concerning science, philosophy, history, society, culture, and education. The conversation takes place between Michel Serres, Bruno Latour, Hans Jörg Rheinberger, Donna Haraway, Sandra Harding, and Hannah Landecker. The group has just attended the opening session for the World Alliance of Science and Science Education Symposium. They make their way to a conference room a short distance from the main auditorium. There is a large overlap in this group of individuals. Some might call them philosophers of science. Others consider them historians of science. Some might say they study the culture of science. Placing these individuals in a concise category is not the purpose of the conversation that needs to take place. The individuals all bring a unique point of view to a discourse involving what science really is and what science should mean to all participants in society, including Henrietta Lacks. The individuals will openly discuss how they view science. Their conversation will allow their views to be shared and compared, as Henrietta Lacks tries to gain access. Not all members of the panel choose to acknowledge the voice of Henrietta as she tries to share her story and convince the panelists that she and her family should be part of science.

Act IV Scene 1: The Conversation Continues

Characters in order of appearance: Hans Jorg Rheinberger*, historian of science; Donna Haraway*, philosopher of science; Sandra Harding*, philosopher of science; Bruno Latour*, philosopher of science; Michel Serres*, philosopher of science; Henrietta Lacks and Mandy Hoffen, teacher/graduate student.

Donna Haraway, Bruno Latour, Sandra Harding, and Michel Serres leave the large auditorium and make their way to a smaller meeting room down the hallway. A man with keys arrives and opens the door, welcoming the group to enter. Everyone takes his or her place around the table. The audience begins to arrive in small groups. A graduate student makes her way to the front row. She holds in her hands a stack of books. Members of the panel recognize them as their own works. The one on top is orange …

The Immortal Life of Henrietta Lacks. This text has been the heart of the discussion so far.

Hans Jörg Rheinberger: Good morning, it is great to see all of you here. I am honored to be sitting at a table with all of you. Thank you Dr. Serres for making this unprecedented appearance. I understand this is a homecoming of sorts for you and Dr. Haraway, both having worked as part of the faculty here at Johns Hopkins. I am most honored to be included as part of this panel. I can begin the introductions. I am Hans Jörg Rheinberger. Feel free to call me Hans. My areas of specialty include philosophy of the biological sciences ... to also include "the history and philosophy of science, in fields ranging from anthropology, sociology, and economics, to literary studies" (Rheinberger, 2010, p. xi). As a scientist and science teacher, I come to you with a rich background in basic scientific experimentation. That might be easier to recognize, had I not checked my lab coat at the front desk. (*The members of the panel chuckle, knowing the stereotypical scientists, working in only laboratories, wearing lab coats, have been long gone—although that was the precedent.*) Hannah Landecker will be joining us shortly. She can help guide the conversation concerning HeLa cells and Henrietta. I know all your work, but I think you should offer a brief synopsis for our audience members.

Donna Haraway: I am Donna Haraway. In my book *Simians, Cyborgs and Women*, I describe myself as a "proper, U.S. socialist, feminist, White feminist hominid biologist, who became a historian of science to write about modern western accounts of monkeys, apes, and women" (Haraway, 1991, p. 1). I am currently exploring "technoscience and the relationships of science to history, culture, and living beings" (Haraway, 1997, p. 25). All of this sounds rather grand and glorious, but honestly, I simply want to be part of the science that can prepare us to live in the 21st century. I have written for years about the treatment of women in science and looking at science with a feminist perspective. I keep thinking about Skloot's remarks about Deborah Lacks. *Haraway sighed deeply.* Deborah and I have something in common. We both lost our mothers at an early age. We both thought we would die of the same disease, at the same age, as our mothers. I was convinced "that I would die when she died, at 42 of a heart attacks, that my body was her body" (Haraway, 2000, p. 14), just as Deborah thought she would die of her mother's cancer. Being preoccupied with dying of cancer is not very logical. In Deborah's case, Henrietta died from a tumor brought about by being exposed to human papillomavirus. And this morning I hear the voices of children about to lose their mother. (*This last comment catches the group by surprise. They look at Donna puzzled. Sandra Harding speaks up quickly, to refocus the group's attention on the required topic of the day.*)

Sandra Harding: Donna, I am so sorry you lost your mother at such a young age. Loss is difficult at any age, but I would think the difficulty would be magnified, even exponential for a young person. (*She looks around hoping it is okay to proceed with her own introduction.*) For those of you who do not know me, I am Sandra Harding. I will also be bringing a feminist point of view to the table today. I want to make a case for what I call "strong objectivity" (Harding, 1991, p. 142). Let me take a moment and explain this. When looking through feminist lenses, objectivity is more stringent than most epistemologies. The feminist "standpoint epistemologies call for recognition of a historical or sociological or cultural relativism—but not for a judgmental or epistemological relativism" (p. 142). This is a "call for the acknowledgment that all human beliefs—including our best scientific beliefs—are socially situated, but they also require a critical evaluation to determine which social situations tend to generate the most objective knowledge claims" (p. 142). The standpoint epistemologies are very different from judgmental relativism—since judgmental relativism does not require "a scientific account of the relationships between historically located belief and maximally objective belief. So, they demand what I shall call strong objectivity in contrast to the weak objectivity of objectivism and its mirror-linked twin, judgmental relativism" (p. 142).

When we hear the story of Henrietta Lacks, the story is very human. All our lives have been touched by cancer. In all honesty, all our lives have been touched by HeLa cells ... do you remember getting your polio vaccine? I ask: Should science be completely objective? I have found that the manner in which objectivism is translated by most will only result in what I call a "semiscience" because it "turns away from the task of critically identifying all those broad, historical social desires, interests, and values that have shaped the agendas, contents and results of the sciences much as they shape the rest of human affairs" (Harding, 1991, p. 143).

Haraway: I have studied this as well. "The natural sciences are legitimately subject to criticisms on the level of 'values,' not just 'facts'" (Haraway, 1989, p. 13). Work in natural sciences "are subject to cultural and political evaluation 'internally' not just 'externally' ... implicated, bound, full of interests and stakes, part of the field of practices that make meanings for real people accounting for situated lives" (p. 13). Scientific observations are not exempt from this process.

Harding: Exactly! We cannot bypass, or sweep under the rug, or wish away what is going wrong in science or our world. We must be encasing ourselves with a true strong objectivity. "The sciences need to legitimate within scientific research as part of practicing science, critical examination of historical values and interests that may be so shared within the scientific community" (Harding, 1991, pp. 146-147). The study of sciences cannot be so segregated by the individual disciplines that "cultural bias" is created "between experimenters or between research communities" (p. 147). Dr. Latour, you are next.

Bruno Latour: I am Bruno Latour. I guess you might call me a French anthropologist, sociologist, and philosopher. I do not find labeling very useful because most categories negate overlaps, pluralities, and endless possibilities. I find that the current categories and boundaries we use in science and science education may be the source of our problems in these fields. We need to become aware that "all of culture and all of nature get churned up again every day" (Latour, 1991/1993, p. 2). I cannot simply read a newspaper each day without noting this phenomenon. I find it disturbing that the stories are presented and seem to say, "let us not mix up knowledge, interest, justice and power ... heaven and earth, the global stage and the local scene, the human and the nonhuman" (p. 3). (*He motions to the gentleman sitting next to him, as if to say next.*)

Michel Serres: I am Michel Serres. Most of you know that I prefer a "solitary exercise of philosophy" (Serres & Latour, 1990/1996, p. 2). I hope I can bring perspective to issues being discussed today that you find foreign—a perspective that is relevant for science of the 21st century. Up to this very moment my work "consisted of preparing for the moment when we would pay the true price of the consequences of science's takeover of all reason, culture, and morality" (p. 86). Science is constantly taking over. If you would like some examples, may I offer Hiroshima and the Holocaust? I have witnessed the attempts of science taking over the world through first-hand experience in the streets of France, where I grew up. Science must have some system of checks and balance. I think philosophy and history can provide this.

Henrietta Lacks: (*sitting on the back row in the shadows*) I have witnessed this takeover of science as well, Dr. Serres. Dr. Rheinberger is not going to introduce me. Why would he? I only have a 6th grade education. I was a wife and mother of five children. I entered the world of white coats downstairs through the same lobby that you entered this morning. However, my Johns Hopkins Hospital was complete with a "colored" bathroom, a "colored" water fountain, and segregated public wards (Skloot, 2010, p. 13). I was just a woman in the colored ward dying of cervical cancer. "I [was] invisible; understand simply because people refuse to see me" (Ellison, 1995, p. 3). To science, I was HeLa—nameless, faceless—just HeLa, not Henrietta.

Haraway makes eye contact with the woman speaking on the back row. Her body jolts, she closes her eyes, shakes her head, and turns to listen to Rheinberger, in hopes of trying to make sense of what was going on with her mind and her eyes this morning.

Rheinberger: Our objective today is to examine current science and science education practices and contemplate the entire HeLa story. Considering the company, I have a feeling ... our discussion will become multidimensional and interdisciplinary! A little history, a little philosophy ... Let us get started. Can I get a volunteer to begin our discussion?

Lacks: This is not just a HeLa story; it is my story too.

Haraway: I would feel better not calling it the HeLa story.

Latour: I will start. I have been ready to go since early this morning. I got here in time to view the HeLa cell art exhibit downstairs. The hall was locked up. Evidently, part of the exhibit was vandalized during the night. I hope they have the exhibit opened later. I find HeLa cells very interesting. I like Sandra's idea that with the separation of fields in science that bias can be created. Would you not say that this type of separation occurs when science is taught in isolation in classrooms? Without the curriculum itself having personal relevance to us, the students of science, and the world that we live in, science has no meaning. How could so many have taught about those miraculous cells without a mere mention of their source? I hope to learn more when I view the HeLa exhibit.

Lacks: (*Henrietta rolls her eyes.*) There he goes again. HeLa … Why won't he use my name? Why won't they say, "Henrietta's cells"? They are my cells, right? They came from my body. Those infamous cells … Call me a "Pyrex monster" (Landecker, 2007, p. 171).

Latour: Science has social implications as well, not all positive. Science education must be more than those standardized tests. I was happy to see in a recent newspaper that parents in the Chicago school district are "opting their kids out of standardized testing" (Wallace, 2015). I tell you that this needs to happen everywhere. I heard that in some districts students are not even participating in the basic science fair projects.

Rheinberger: I always thought those science fair projects were a little narrow minded … Maybe narrow minded is not a good word. Let me try this … I think the standardized structure gives students an idea that all science comes about in nice neat steps … but in lieu of standardization, projects would be better than students completing giant stacks of test practice worksheets. Why is it that people always attach the idea of a strict experiment to science?

Harding: There are so many more possibilities that could come to life in science class. There is a teacher presenting later today. Did you see it on the program? She uses the story of Henrietta Lacks and HeLa cells to teach high school biology. Her abstract says the students respond to the story by writing songs, poetry, art, and letters to Henrietta herself or to her family members.

Lacks: Finally, someone said Henrietta's story.

Rheinberger: I wonder if she lets them look at HeLa cells under a microscope. In science class, one would not expect to hear about poetry and art. Science class is usually dominated by the strict scientific method. I have always found it interesting that "the experimental system is the smallest functional unit of science" and yet it gets the most press (Rheinberger, 1992, p.

306). This small unit is "designed to give answers to questions which we are not yet able clearly to ask. It is a machine for making the future" (p. 309). The experimental unit "is not only a device that generates answers: at the same time and as a prerequisite, it shapes the questions that are going to be answered ... a device to materialize questions" (p. 309). This process of experimentation is not unlike "any particular auto-biographical narrative" (p. 310). What some may not realize is that these small experimental data units are much more than one would think. These units connected may indeed tell a story.

I would personally like to see science consider experimental systems and move in a direction where the "kind of movement in which reasoning makes itself by way of tracing" (Rheinberger, 1992, p. 306). Tracing is a natural progression in which the pathway is not predetermined. Actions are determined based on need, not a prescribed, predetermined method. The scientist moves about the information unrehearsed, unscripted. One might call it laboratory improvisation. We could make a reality show ... sorry for my digression from the formal conversation this morning. Let's think about Paul Feyerabend's work for a moment (1975/2010). He tells us that "theories become clear and 'reasonable' only after incoherent parts of them have been used for a long time. Such unreasonable, nonsensical, unmethodical foreplay thus turns out to be an unavoidable precondition of clarity and empirical success" (Feyerabend, 1975/2010, p. 11). In this somewhat "concrete, practical process of the search for knowledge, explicit methodological rules are rather counterproductive, only causing additional confusion where they are supposed to produce clarity" (Rheinberger, 2010, p. 63). Feyerabend (1975/2010) "felt that the history of science after all does not just consist of facts and conclusions drawn from facts. It also contains ideas, interpretations of facts, problems created by conflicting interpretations, mistakes and so on" (p. 3).

I will continue by saying that:

> Instead of speaking about theories, experiments, instruments, and their connection in order to characterize scientific enterprise ... I speak about the experimental situation, about scientific objects or 'epistemic things,' the differential reproduction of experimental systems, and the conjuncture of each system. (Rheinberger, 1992, p. 307)

This teacher you mentioned, if she is having students learn about HeLa cells from the story of Henrietta Lacks, is she tracing the history of cell lines? With all the implications of this story, tracing the history would allow the students to consider so much more than a basic experiment.

Harding: There must be a tracing of some sort … think about the knowledge generated from scientific research. Once an experiment has ended and the data is collected, new knowledge produced, students are taught that the process is repeated, so the new knowledge can be confirmed. Real science we know does not end with simple knowledge production. There will be consequences to accompany the new knowledge. Consequences that affect society. Consequences that should be analyzed before, during, and after experimenting. This analysis should be very detailed and multidimensional. Without this detailed analysis, scientists are practicing what I call "weak objectivity" (Harding, 1991, p. 147). Weak objectivity creates a situation where "scientists produce claims that will be regarded as objectively valid without their having to examine critically their own historical commitments, from which—intentionally or not—they construct their scientific research" (p. 147). Allowing the practice of weak objectivity in science "permits scientists and science institutions to be unconcerned with the origins of consequences of their problematics and practices, or with the social values and interests that these practices support" (p. 147).

Lacks: Hmm … Weak objectivity. Does this mean that scientists think that my children being able to go to school and tell the other children that their mom's cells "wrap around the earth at least three times, spanning more than 350 million feet" is pretty cool? I would not mind that. I do not mind them sharing that my cells helped millions of people … but there is more to the story. Science should take care of people in need. I would say there are cases where the title human experimentation hides human exploitation. When my battle with cervical cancer began, Johns Hopkins was the only hospital where I could go for treatment. Richard Wesley TeLinde was one of the top cervical cancer experts in the country; he justified the use of the innocent for the advancement of science … for the good of all humanity. Howard Jones, my doctor, wrote, "Hopkins, with its large indigent Black population, had no dearth of clinical material" (Skloot, 2010, p. 30). Students of science need to know the circumstances that landed my cells on the moon. Then maybe they will ask questions like: Did the doctors give Henrietta the same dosage of radium that any other woman would have received? Was Henrietta being treated for her cancer, or was she just another experiment, making payment for her debt ... her room and board in the public ward?

Haraway: Room and board in the public ward (*she repeated under her breath*), while her children looked up at her window from the park below. Excuse me Hans, did you hear that? (*She looks around, shakes her head, and reaches for her coffee. Was the voice she heard a member of the audience?*) Please excuse me for interrupting. (*She scans the meeting hall looking for a face that might have matched the voice. The voice seems to be coming from the shadowy back row.*)

Rheinberger: No problem. I did not hear anything. All these systems are related to the "practical process of producing knowledge" (Rheinberger, 1992, p. 307). I do not want to deny that the experiment is important to research—it may serve as the foundation of research. However, it is only one component, "the smallest functional unit of research" (p. 309). If we start a list as we go, we can keep up with what we think science and science education need to mean to students of science ... at all levels. Can we all agree that working in laboratories to create knowledge is important? I think we have covered that the strict method is not necessary, but the basic process of inquiry should be taught and understood by everyone. (*Rheinberger writes "Basic laboratory skills of inquiry/not-strict method" beside number one on his notes.*)

Harding: I agree Hans; it is the experimenting that produces what the world considers science or scientific knowledge.

> Science has been held in such a high regard since Sputnik, of course—indeed, ever since Newton—but the flood of industrial and federal funds that pours into scientific and technological projects in universities these days is truly astounding. It is a long time since scientific research could be regarded as significantly isolated in real life from the goals of the state and industry—if it ever could. Scientific research is an important part of the economic base of modern Western societies. (Harding, 1991, p. 4)

Lacks: (*Louder this time from the back row.*) I hear from your conversation that you have a high regard for the process of experimenting. Do you and your students of science understand the implications of using humans for experimentation without their consent? Have you all forgotten about the Tuskegee syphilis experiment? How about all the people that died at the Crownsville Hospital? What about the prisoners that were injected with my cancerous HeLa cells?

Haraway: No, I have not forgotten about Tuskegee. Yes (*speaking to address the woman on the back row*), I do understand the implications of humans being exploited. I am appalled that those researchers withheld treatment. Those men died horrible, painful deaths. (*She stands and speaks in Henrietta's direction; then realizes she has interrupted again. It is obvious she is the only one who can see the woman on the back row, much less hear her. She sits down quickly.*)

Harding: Donna, I was talking about the economic connection to science. Not exploitation, I think. Did I say exploitation instead of economics?

Haraway: We have someone on the back row who wants to join in on the conversation. I was responding to her. The woman on the back row, she is wearing a brown suit. She mentioned the Tuskegee syphilis experiment. She also mentioned the Crownsville Hospital. I am not sure I know about the Crownsville hospital.

Rheinberger: I think you both are right. Economics of science can lead to exploitation in science. Go back to 1951 in this very hospital. The public wards were full of individuals who were not able to afford health insurance. Many individuals because of their race could not even gain admittance to hospitals ... they might have just "died in the parking lot" (Skloot, 2010, p. 15). Yesterday, in the exhibit downstairs, I read that Johns Hopkins Hospital was one of the only hospitals that allowed African Americans to be admitted to the hospital's public wards.

Haraway: Henrietta, right? Did you have children? Deborah was the baby?

Lacks: Yes, I had five children, before I was 30. My youngest son, Joseph, was born right here in this hospital. Deborah was 3 when I died.

Haraway: Wow, five! When I was much younger, with my Catholic upbringing, "I figured I'd have 10 children" (Haraway, 2000, p. 10). (*Donna chuckles, and then realizes she is getting loud. Harding thinks she is asking a question and jumps in to answer it.*)

Harding: She was a mother of five, just 30 years old when George Gey took her cells. I read about it in a display case downstairs this morning.

Latour: (*Using his tablet to read from the conference program.*) It says here that Crownsville was a hospital for the Negro insane. There is an excerpt posted from Rebecca Skloot's book, *The Immortal Life of Henrietta Lacks*. You can find it in the back of the program catalogue, in its own appendix. A representative for the hospital told Skloot and Deborah that, "there wasn't much funding for treating Blacks in the forties and fifties ... and Crownsville wasn't a very nice place to be back then" (Skloot, 2010, p. 270). The two were visiting to retrieve medical records for Elsie, Henrietta's daughter who died at Crownsville.

Lacks: My child died of "(a) respiratory failure (b) epilepsy (c) cerebral palsy ... she spent 5 years in Crownsville State Hospital" (Skloot, 2010, p. 270).

Haraway: That is horrible, Henrietta, I am so, so, sorry. (*The panel members all turn and stare at Donna Haraway. Rheinberger continues.*)

Rheinberger: Unfortunately, no money for the poor is not a new problem.

Haraway: I mean your daughter. Her Daughter. You died here; your daughter died five years later in a place of unspeakable horror all in the name of medical science. (*The panel is starting to wonder about Donna Haraway. She notices they are staring at her.*)

Serres: (*He has been reading in the program during the entire conversation.*) It says here that Crownsville hospital had "a serious asbestos problem" (Skloot, 2010, p. 275). Again, as Sandra said ... a problem of economics. Money for war, but no money to take care of the sick and needy at home. It says here that in 1955, patients "arrived from a nearby institution packed in a train car ... the population was at a record high of more than 2700 patients, nearly eight hundred above maximum capacity" (p. 275).

Harding: This was a horrible place (*reading from her iPad*).

> Crownsville averaged one doctor for every 225 patients, and its death rate was far higher than its discharge rates. Patients were locked in poorly ventilated cell blocks with drains on the floors instead of toilets. Black men, women, and children suffering with everything from dementia and tuberculosis to nervousness, lack of self-confidence, and epilepsy were packed into every conceivable space, including windowless basement rooms and barred-in porches. (Skloot, 2010, p. 275)

Haraway: We also must think about money for biotechnology. Think about the economics of HeLa cells for a moment. Scientists go into laboratories and come out with living and breathing commodities—attached to real people. "Living organisms can become patentable 'composition of matter'" (Haraway, 1997, p. 90). This is when science gets complicated. Are the HeLa cells considered a living organism, attached to Henrietta? Who owns the cells—they came from Henrietta, but she died, so is ownership passed to her family? In this morning's discussion, it was agreed that George Gey had no desire to make HeLa cells a commodity. Gey wanted to save the world from cancer. However, just because the cells are never patented does not mean that the cells did not become a commodity.

Rheinberger: Should we add as # 2—Science connections to real life, including economics? (*Everyone nods in agreement, and sees that Rheinberger, as session leader, has been charged with keeping the discussion on track.*)

Latour: I want to make sure students of science realize that science is connected to culture. As students of science, we are charged to learn these long lists of facts. These lists are meaningless if students do not see the connections to culture. Take a glance at a basic curriculum map given to a high school science teacher … If this list is indeed an indicator of what is known about science and what should be taught about science, it becomes obvious that science "knows nothing about culture" (Serres & Latour, 1990/1996, p. 86). "Science has all the power, all the knowledge, all the rationality, all the rights too of course, all plausibility or legitimacy, admittedly—but at the same time all the problems and soon all the responsibilities" (Serres & Latour, 1990/1996, p. 87). Science is looked upon as an all-knowing discipline on "top of the heap, as we say, single-handedly preparing the future" (p. 87). With this scenario, our future is grave.

Serres: "As science advances, we rarely evaluate the substantial cultural losses that correspond to the gains" (Serres & Latour, 1990/1996, pp. 54-55). Bombings, war, no thoughts of the personal devastation to follow!

Latour: It is obvious that "All scientists can sketch out a brief history in which they place themselves at the pinnacle of reason, after centuries of groping" (Serres & Latour, 1990/1996, p. 51). Michel, your perspective of science is connected to the war, the bombings. There must be connections—an anchor of sorts— context. Back to the comment about never having heard Henrietta's name in reference to HeLa cells. Teaching HeLa without teaching the entire story of the cells is a missed opportunity for educators. There is so much history in the story, history I believe every medical student needs to know, but more importantly, details every patient needs to know, such as how they should make sure consent forms are read and understood.

Haraway: I need to back up for just a moment. Maybe I am being too analytical and over- sensitive this morning. (*It has been a strange morning; she thinks to herself.*) Dr. Serres, didn't Einstein try to inform the president about nuclear weapons (Einstein, 1929)? I think reflection is very important in these complex matters, constant reflection—not sporadic or because one has made it to the end of a prescribed number of steps. We must be committed to "science-in-the-making" and the making of a science that will cease to "reproduce systems of stratified inequality—and that issue in the protean, historically specific, marked bodies of race, sex and class—for performed, functionalist categories" (Haraway, 1997, p. 36). Is it not new to link the stories of science and democracy, "anymore that it is to link science, genius, and art, or to link strange night births and manly scientific creations? However, the interlocking family of narratives in the contemporary U.S. technoscience drama is stunning" (p. 168). The more I think about it today, the more passionate I feel about the need for more humanity in science.

Rheinberger: I am trying to keep up here … #3 Science connections to culture.

Latour: Donna, I think your idea of crossing boundaries is what technoscience infers. The interdisciplinary contributions of the philosophy, the history, the culture of our world, all must be a part of science. I am not sure if incorporating human or humanity is necessary, I think it is understood.

Haraway: I think we must be careful not to disregard any possibilities. We must work hard to "maintain this very potent joint between fact and fiction, between the literal and the figurative or tropic, between the scientific and the expressive" and between Henrietta and HeLa (Haraway, 2000, p. 50). Maybe I am overstating here … but humanity is our priority. "Interdisciplinarity is risky but how else are new things going to be nurtured?" (p. 46).

Rheinberger: I will add #4—Science connections to humanity. (*The others are not very concerned with Rheinberger's list. They are all intensely wrapped up in the discourse.*)

Latour: Current boundaries in science bog us down. If science is only being viewed through a lens focused on strict experimentation, we are neglecting a world of possibilities. Think about that simple morning spent with your cup of coffee and the morning paper. The day's stories are about "global warming, AIDS vaccines, frozen embryos, the Pope and contraception pills" (Doll et al., 2001, p. 25). These stories follow a complicated path such as this one I wrote about in *We Have Never Been Modern*:

> The smallest AIDS virus takes you from sex to the unconscious, then to Africa, tissue cultures, DNA and San Francisco, but the analysts, thinkers, journalists and decision-makers will slice the delicate network traced by the virus for you into tidy compartments where you only find science, only economy, only social phenomena, only local news, only sentiment, only sex. Press the most innocent aerosol button and you'll be heading for the Antarctic and from there to the University of California at Irvine, the mountain ranges of Lyon, the chemistry of inert gases, and then maybe the United Nations … They seem to say let us not mix up knowledge, interest, justice and power. Let us not mix up heaven and earth, the global stage and the local scene, the human and the non-human. (Latour, 1987, pp. 2-3)

When you see all of this … even though your mind may be fighting to categorize everything into nice little cubicles … it is obvious the named disciplines are mixed. These mixtures "weave our world together" (Latour, 1993, p. 3). This morning all my connections were made through HeLa cells.

(Everyone on the panel simultaneously says Henrietta Lacks in chorus.)

Latour: Point taken. *(He glances to the back of the audience, and for a moment he thinks he sees someone who resembles that classic figure in the brown suit.)*

Rheinberger: A woven pattern, a spider web, rhizomes—yes, Bruno, this is exactly what I was speaking of earlier. Just like newspapers, science tries to put everything into nice little labeled boxes with predetermined categories allowing for no overlap; the networks that are created by tracing events, in multiple ways. It is no doubt that real events of the world will need to enter this web. So, #5—Interdisciplinarity—science connections to other subjects ... history, humanities, mathematics, anthropology, sociology, political science. Help me. What else should I add?

Harding: We cannot make an all-inclusive list. The possibilities are endless.

Haraway: *(appears to be snapping out of a daze, she adds quickly, making eye contact with Rheinberger)* Zoology, philosophy, literature, art, music ...

Latour: Let me place our conversation in the context of hybrids. I have always defined hybrids as being composed of two competing sides but remember that each side does not have to be a single entity. If so, then it would just be another binary—I think that is what we all want to avoid. I do not see hybrids as a combination of two things, like a mule, which is the cross between a horse and a donkey, or a pluot.

Rheinberger: A pluot?

Latour: A plum crossed with an apricot *(he says quickly)*. "On the left, they have put knowledge of things; on the right, power and human politics" (Latour, 1991/1993, p. 3). Our world views knowledge as an isolated entity, separate and removed from power and politics. The junction that I desire to create shows that science can be connected to people, politics, and policy. This junction provides us with further reasoning that science must be continually in the making (Latour, 1987).

Lacks: *(Addressing Donna)* Hybrids seem to connect everything. But when I hear the word hybrid, I think about those crazy picture shows my sisters and I would go see on Saturday afternoons—

maybe, *Dr. Jekyll and Mr. Hyde*. Sounds like that is how scientists did and are describing me. Homemaker that saved the world … and cells that destroyed research, labs, and even threatened world peace. I want to know if these connections are just for the scientists. My experience with science was situated on the wrong end of the binary … no hybrid … no connection. Just a poor, Black woman in the public ward leaving behind five kids.

Haraway: Henrietta, you are right. The term hybrid is supposed to show a bringing together of ideas. They should help how people think … but if a hybrid only brings about two ideas … we just have another binary, like Dr. Jekyll and Mr. Hyde. Scientists have personified the personality of your cells onto you as a person.

Hannah Landecker enters the room and glances across the audience. She stops when she sees Henrietta. She looks away thinking, I must be seeing things, and addresses the panelists on the stage.

Hannah Landecker: I apologize for missing your opening. I was assisting with the police report concerning the vandalism that occurred in the exhibit hall. (*Noticing the puzzled faces, she explains further.*) The BioArt exhibit was knocked over and the mannequin's shoes, suit, and earrings are missing. (*She sits down in the empty chair by Donna Haraway.*) We are lucky the whole building did not burn down. Supposedly, there was an electrical short in the pink Henrietta sign. Did I hear someone mention Jekyll and Hyde?

Harding: I cannot believe someone would vandalize the display. Did you all see the display of cells in the lobby? I thought it was very interesting to include living cells in the display. I have not looked through a microscope in years. The cells are very interesting to look at, especially when you simultaneously think about the story. Skloot's book, *The Immortal Life of Henrietta Lacks*, brings science and culture together. She has touched a huge audience base. I like how the artist in the exhibit included pictures and parts of Henrietta's story in the display. That does bring home the point that Henrietta's cells grown in culture came from a real woman … and these cells did affect humanity. (*She whispers to Haraway.*) That woman on the back-row kind of looks like Henrietta.

Haraway: That is whom I was speaking to earlier. I thought for a moment that I was the only one seeing her and hearing her.

Latour: Maybe it is time for those new glasses. (*He squints toward the audience.*)

Serres: Sometimes we see what we want to see. If we cannot envision a changing world, we will never see changes take place in the world.

Harding: Change: That is the most important topic of today.

Rheinberger: Oh, we do need to put that science is constantly changing on this list. What are we on, #6? (*He writes* Science continually changing.)

Landecker: Hans, thanks so much for following the directions. I am so happy you are indeed composing a list of what science and science education should include.

Harding: I often wonder "how should philosophies of science be reshaped to account for modern sciences' history, achievements, limitations and possible futures identified in studies, that share skepticism about the conventional internalist epistemologies of science"? (Harding, 1998, p. 5). I think the "convergences and divergences" as well as "feminist components" could bring about revelations to challenge current knowledge traditions (p. 5).

Latour: So, a multicultural, feminist point of view ... hmm ... all connected to history, economics and society.

Rheinberger: #7 Science from a feminist perspective.

Landecker: I commented this morning on how the story of Henrietta Lacks was portrayed from a racist point of view. How would her story be different if it had been told from a multicultural and feminist perspective?

Haraway: Well for starters, I think it is obvious; there would be no racialized language such as aggressive, promiscuous ... and the constant use of the word, "she." A cell is a cell; they are not gendered until humans start with the anthropomorphism!

Harding: When telling this story, "cultures and practices of the sciences should be understood to provide the necessary conditions for sciences to do their work, but they should not influence the results of research in any culturally distinctive way. All social values and interests that might initially get into the results of scientific research should be weeded out as soon as possible through subsequent critical vigilance" (Harding, 1998, p. 3). The public ward provided a predisposition of cultural bias, as was common practice in those days of segregation.

Rheinberger: Culture does have a role, but cultural bias does not. When I think about this story, I look at it from a laboratory point of view. Hindsight for George Gey would be 20/20. "There is no prespecifiable golden path or universal method that can provide a blueprint for knowledge production" (Rheinberger, 2010, p. xvi). We have spoken extensively about the experimental unit today. We know that method can only provide so much assistance in a very chaotic environment. "Research is inherently untidy and requires an appropriate cultural setting that sustains innovation by allowing the new and unpredictable to emerge. Environments that sustain controlled chaos are the heartland of innovation" (p. xvi). A problem in a laboratory is like a problem anywhere else in the world, there should be lots of discussion with a thorough examination of all possible perspectives. I think there probably was lots of discussion in the early days of HeLa cells research, but did anyone think of the story from Henrietta's viewpoint? The perspective of her family? Did scientists consider the future possibilities and implications for science? Problems viewed in isolation will most likely be addressed in a very narrow-minded way. This story is interdisciplinary. Chaos was created from the moment; Kubicek had to figure out a way to store all those cells. Gey had to figure out what to do with them. When chaos is created, the goal is to control it. "Environments that sustain controlled chaos are the heartland of innovation" (p. xvi). I guess what I am saying is that all the perspectives would have something to offer ... leaving out other tellings of the story narrows the outcome. Did they leave out Henrietta's name for her own protection?

Serres: (*He has continued to read while listening to the conversation.*) I think everyone in the lab knew exactly from whom the cells came. It says here in Skloot's book "*The Minneapolis Star* became the first publication to name the woman behind the HeLa cells ... but the reporter got her name wrong" (Skloot, 2010, p. 105). "Keeping patient information private was emerging as a standard practice, but it wasn't law" (Skloot, 2010, p. 107).

Haraway: I think people might see Henrietta's role differently if I share with you the story of the OncoMouse. The OncoMouse is "a breast cancer research model produced by genetic engineering" (Haraway, 1997, p. 47)—it is a genetically modified mouse that carries a specific cancer-causing oncogene. "As a model, the transgenic mouse is both a trope and a tool that reconfigures biological knowledge, laboratory practice, property law, economic fortunes and collective personal hopes and fears" (p. 47). Can you flip on that projector for me, Hans? The painting shown depicts the white female mouse, complete with breasts and a crown of thorns. "She is a Christ figure, and her story is that of the Passion ... The laboratory animal is sacrificed; her suffering promises to relieve our own; she is a scapegoat and a surrogate" (47). I see the chamber as comparable to the cubicle chamber and to a box that "mimes the chamber of the air-pump in Robert Boyle's house in 17th century England (p. 47). In this chamber in Robert Boyle's laboratory, "small animals expired ... to show to credible witnesses the workings of the vacuum air-pump so that contingent matters of fact might ground less deadly than social orders" (p. 47). The mouse in present day "is a figure in secularized Christian salvation history and in the linked narratives of the Scientific Revolution and the New World Order—with their promises of progress; cures; profit; and if not of eternal life, then at least of life itself (p. 47). Science is all these things. Science promises us solutions to current problems. Science promises us cures from diseases with gene therapy. The OncoMouse has witnessed it all. When we speak of curriculum, OncoMouse verifies for me that Science should be interdisciplinary. The history and the story behind the OncoMouse are evidence that Science needs the humanities in order to tell a story that the world finds relevant. Science should know no bias to gender or race, Science should simply know humanity and all that humanity encompasses.

Let us take a moment and return to 1951, right here at Johns Hopkins. It is not OncoMouse or Boyle's small rodents. George Gey, Howard Jones, and George TeLinde have placed Henrietta Lacks in a box of sorts. Henrietta Lacks becomes the object of circumstances. Her box is the public ward, then the basement laboratory. No one is looking down on Henrietta, because no one knows who she is. The world is looking at

HeLa cells to be the savior, but the idea that these cells are connected to their donor is only presented through stereotypical racialized language. Henrietta herself is invisible. Take this story one step further to Crownsville. Again, there are no windows—no one to see Elsie or the others who died needlessly. The men and women that arrived healthy died because there were no windows, no eyes looking on—no one cared. The feminist version of this story spares no details. The story if viewed through multiple lenses will give rise to multiple versions. In the feminist version, Henrietta would be part of the HeLa story.

Harding: Why consider science from the view of gender difference? This perspective "leads us to ask questions about nature, and social relations from the perspective of devalued and neglected lives. Doing so begins research in the perspective from the lives of 'strangers' … It starts research in the perspective from the lives of the systematically oppressed, exploited, and dominated" (Harding, 1991, p. 150). I can go back to strong objectivity here, the feminist point of view "requires a commitment to acknowledge the historical character of every belief or set of beliefs—a commitment to cultural, sociological, historical relativism" (p. 156).

Haraway: I want to make sure the audience understands that feminist science is not just science for women; it is a science that is all-inclusive of those continually left out. This point of view will take on issues involving the fact that stereotypical laboratory or experimental life "required a special, bounded community" (Haraway, 1997, p. 25). We must try to begin "restructuring that space [which]—materially and epistemologically—is very much at the heart of late-twentieth-century reconsiderations of what will count as the best science" (p. 26). That restructured community will need to include scientists, teachers, everyday people, Henrietta Lacks, and others.

Rheinberger looks at his watch … there is only 15 minutes left in their session.

Rheinberger: How about I record for #8—Science is not just for scientists; it must include all participants? Not to change gears abruptly, but there is one more aspect of this discussion I would like us to consider. I want us to go back to why science must be situated in what is going on with society. Have you ever thought about the role of science in "Industrialization, the saturation propaganda of governments and advertisers, two world wars, the concentration camps, the dimming of faith by Science, and of course the constant threat of nuclear annihilation" (Wolff, 2003, p. 52)? Science brought about advances that made industrialization possible. Propaganda and advertisers would be helpless without the technology science brought into being. Weapons for war, methods for torture changed lives forever.

Harding: Think about industrialization from a Western point of view. Our "modern Western sciences and their technologies have always been regarded with both enthusiasm and dread" (Harding, 1991, p. 2). Giving credit where credit is due, we owe "at least some responsibility for the high standard of living that many in the West enjoy—especially if we are White and middle or upper class" to modern sciences and their technologies (p. 2). I don't think any one of us sitting here today would "want to give up the food and clothing, medical treatment, cars and airplanes, computers, television sets, and telephones that have become available through scientific and technological development" (p. 2). On the flip side of the coin, "just who or what is responsible for atomic bombs, Agent Orange, industrial exploitation, polluted air, and vast oil spills, dangerous contraceptives such as Dalkon shields, inappropriate uses of Valium, health profiteering, and high infant mortality in the United States?" (p. 2). We could add "famine in Ethiopia, and the development of a Black underclass in the United States" to the list (p. 2). Science is willing to take full credit "for the good aspects of the Western way of life" but "misuses and abuses are entirely the fault either of politicians or of the industries that apply supposedly pure information in socially irresponsible ways" (p. 2). In Mary Shelley's (1990) dystopian novel, the scientist's name, Frankenstein, "in popular thought migrated to the monster that he [Frankenstein] inadvertently created" (p. 2). The monster is "created—gets nourished and reproduced day after day—retreats into the shadows" (p. 2). The monster becomes part of the shadows ... the place where "there are no persons or institutional practices that we can hold responsible for the shape of the sciences and the kind of social order with which they have been in partnership" (p. 2).

Landecker: The Frankenstein connection is very important. It is very easy for science not to be held responsible. Think about Henrietta. It was very easy to keep her cells growing in the shadows, especially before her name was released to the public. The doctors took her cells without permission … but it was "okay" because it was the practice of the day. Her cells were sent all over the world, but no one ever paid for the cells. Did George Gey, ever consider "what scientific, legal, economic, and rhetorical practices maintain the condition of their existence" (Landecker, 1999, p. 205)? It is almost like the cart before the horse problem, create the science, and then find a use for it. Dr. Serres, do you think scientists refused to take responsibility for the creation of the bomb?

Serres: "As a child of the war and its bombings formed by the horror of the concentration camps, I have always preferred to construct or put together rather than destroy" (Serres & Latour, 1990/1996, p. 26). "I love and seek peace which seems to me the ultimate good" (p. 24). Even with my efforts toward seeing the ultimate good, the tragedy brought about by science cannot be ignored. "As soon as inventions or scientific results and projects pose redoubtable global questions, touching people's lives or the survival of the globe, we hear the cry 'Let's form ethics committees; let's bring in the legal profession, philosophers, and the clergy'" (Serres & Latour, 1990/1996, p. 86). Ethics committees were formed after the fact. We cannot forget uncomfortable parts of our history such as Hiroshima. "Hiroshima was truly the end of one world and the beginning of a new adventure. Science had just gained such power that it could virtually destroy the planet" (p. 87). Hannah Arendt (1951) tells us that "we can no longer afford to take that which was good in the past and simply call it our heritage, to discard the bad and simply think of it as a dead load which by itself time will bury in oblivion" (p. ix). On one hand Hiroshima was a horrible tragedy initiated by science and power. On the other hand, some felt it was necessary to save the world from war and save lives.

> The first atomic bombs were dropped on Hiroshima and Nagasaki on the 6th and the 9th of August 1945, and they presented the ideal conditions: great mechanical effectiveness, complete technical surprise, but above all, the moral shock that suddenly banished to the prop-room the earlier strategic carpet-bombing of large Asian and European cities, with all its logistical sluggishness. By demonstrating that they would not recoil from a civilian holocaust, the Americans triggered in the minds of the enemy that information explosion which Einstein, toward the

end of his life, thought to be as formidable as the atomic blast itself. (Virilio, 1989, p. 8)

This brings to life memories of a horrible event in history brought about by science. The Manhattan Project brought together top scientists to develop weapons of mass destruction. The destruction was witnessed by the world … yet the world did not learn very much. The information relayed that life itself is less important than the power these weapons provide. We need to keep those memories alive so history will not repeat itself. Unfortunately, "time heals, but it does so through erasure" (Rowlands, 2009, p. 4).

Haraway: At the Los Alamos National Laboratories in New Mexico, there is an exhibit about the "first atomic bombs built during the Manhattan Project" (Haraway, 1997, p. 53). The exhibit is "rather mouse nibbled and time-worn; it looked like old news" (p. 53). "The glitzier projects" such as the "Human Genome Project" and the "artificial life research" appear to take precedence over the H-bomb (p. 53). "Nuclear weapon research—albeit still quite a growing concern—is almost, but not quite, an embarrassment even at the birthplace of the atomic bomb" (p. 53). The Human Genome Project has been greeted with wonder and expectations of curing diseases of all forms.

Serres: Look at how our biosphere has progressed over the last 50 years. We can micro-engineer body implants and procreate organisms for the sole purpose of providing organ replacements. The "bricolage of transgenic monsters, cloning of cells, etc.—should be greeted with a sense of awe, even if one doesn't believe that the human body had necessarily reached its final stage" (Virilio & Lotringer, 2002, p. 17).They call this an "assault on the human race … too reminiscent of ethno-genocidal horrors for anyone to trust the neutrality of science" (p. 17). I can relate to Virilio's telling of this story. I did not experience the carefree youth that most of our privileged students of science enjoy. My generation was impacted and traumatized by "violence, death, blood and tears, hunger, bombings, and deportations" (Serres & Latour, 1990/1996, p. 2). "Now that I am older, I am still hungry with the same famine, I still hear the same sirens; I would feel sick at the same violence, to my dying day" (p. 4). The horrors I witnessed shaped my very being. Science created the weapons of mass destruction that created the violence that my generation endured. I have built a life searching for peace. In order to find peace, I found myself creating, synthesizing, and inventing—my own self-taught style of philosophy. "When a person's life begins with the experience and atmosphere of death, it can only move forward in an ongoing spirit of birth, of rebirth, of a positive and overflowing wellspring of exhilaration" (p. 42).

Lacks: My children did not experience a carefree youth either. Mother burned from inside out with radiation … sister dead in hospital for the Negro Insane … children abused following mother's death. Carefree youth was something my children never knew either, Dr. Serres.

Serres: Human, collective, political, economic, social conditions— poverty for example—prevail, and by far, over the purely physical cause "How is it that the lights we receive from the sciences are sometimes accompanied by such blindnesses? The better we know, the more we can. How to go from these possibilities to the real without posing the problem of evil? What responsibility did the scientists of the Manhattan Project formerly bear?" (Serres, 2010/2012, p. 58).

Lacks: I know I can never equate what happened with me to the horrors of Hiroshima. Still, I cannot help but think that the scientists had some responsibility in my story as well.

Serres: Let us go back and hover over the past for a moment. "Supposing that we had foreseen everything, what would we have done? One does not stop science or progress, even less the necessity to win a just war" (Serres & Latour, 1990/1996, p. 59). Is it an issue of the "libido of the dominant males that always drives them to show themselves to be the strongest of all, the fierce competition, and the passion of small brains to arrive first" (p. 59)? A simple contest? A sporting event of sorts? Think about the Manhattan Project celebrities, "vainglorious and appearance-conscious like prima donnas, wanting to succeed at all costs, before the Nazi physicists" (Serres, 2010/2012, p. 59). Going back and looking at the past is only productive if we consider the lessons of the past, in reference to issues that need resolution in the present and future. "If yesterday's truth is tomorrow's error, then in the sciences it likewise happens that the error condemned today will sooner or later find itself in the house of great discoveries" (pp. 52–53). Past or present discourse is a necessary element of science for the examination of such issues.

Rheinberger: I am adding discourse in science as # 9. I think that is an obvious addition. Discourse will be necessary to solve the modern-day problems involving science.

Haraway: Science is the result of located practices at all levels" (Haraway, 1997, p. 36). "The point is to make a difference in the world, to cast our lot for some ways of life and not others. To do that one must be in action, be finite and dirty, not transcendent and clean" (p. 36). "We must, however, be acutely aware of the dangers of using old rules to tell new tales" (Haraway, 1991, p. 42). Science can be viewed as a process that is "resolving the contradiction between or the gap between human reality and human possibility in history" (p. 42).

Serres: "Caused, causing, all things in the world ensue from each other, chained together" (Serres, 2010/2012, p. 129). "Whether fluid or of air even solids communicate things respire together, they conspire with different breaths, but in constant and total circulation that's chancy, torn, chaotic and consenting" (p. 129).

The conversation ends abruptly.

Rheinberger: Unfortunately, we are out of time. Do we have any quick questions or comments from the audience? (*The restless audience begins to leave.*)

Hoffen: (*sitting on the front row*) I have one question. Can Henrietta Lacks become part of this new science?

Henrietta begins walking toward the stage, as others are leaving the room, some glance at her, others just walk by, unaware of her presence.

Rheinberger: I noticed you have Rebecca Skloot's book. It looks as if it is worn from a lot of study. What do you think? Here is the brief shopping list of sorts we created today concerning what elements are necessary for science to be meaningful to all students of science. To all participants of science, this list needs to be proofed, revised, and maybe even reordered ... but what do you think?

1.   Basic laboratory skills of inquiry/not strict method;
2.   Science connections to real life, including economics;
3.   Science connections to culture;

4. Science connections to humanity;

5. Science connections to other subjects (Interdisciplinarity ... history, humanities, mathematics, anthropology, sociology, political science, and more);

6. Science connections to constant change;

7. Science from a feminist perspective;

8. Science is not just for scientists, it must include all participants;

9. Science must incorporate Discourse in order to solve problems; and

10.

Oh no ... I thought we had 10. Hannah wanted 10.

Hannah Landecker: Hans (*shaking her head and smiling*). You know I need 10.

Hoffen: (*interrupting*) I have a very difficult time separating Henrietta from her cells. We as human beings are connected to this story of Henrietta Lacks through the antibodies of poliovirus that pulse through our veins. We are connected anonymously, until the moment we hear Henrietta's story. The polio vaccine was tested on HeLa cells. Henrietta was a living, breathing woman struck down by what started as a microscopic tumor-producing virus, HPV (human papillomavirus), and grew like rhizomes to overtake her body. Her name and face were both lost in the public wards of Johns Hopkins Hospital. The poliovirus was received first by the privileged White scientists. This vaccine was made with a poor Black woman's cancer cells. Her family left behind to take care of themselves in an unbroken cycle of poverty that continues to marginalize and reduce opportunities for those of specific genders, race, and class. (*Henrietta walks right up beside the group. Hoffen glances back and jumps. Is what she is seeing real? Some audience members turn and see the beautiful African American woman, dressed in a brown suit standing in the front with the panel. Some gasp at the resemblance between her and the mannequin they had observed the day before in the exhibit hall. Others continue to walk out, not seeming to see her. The panelists have their eyes and ears glued as she begins to speak.*)

Lacks: You are telling my story. Can science hear my voice? Or am I simply a dead branch wavering in the wind? A ghost—visible to only a small few, who believe the "universe is composed of stories" (Rukeyser, 1968, p. 11)? Those who know me as Henrietta, not just HeLa.

Serres: (*making eye contact with Henrietta*) Yes, some can hear your voice and see you here today. Others will not open their minds to hear nor see you. It is interesting, "the sciences are beginning to discover ... the trees themselves emit voices. Who chatters in this concert? Who speaks in total...I wager that tomorrow advanced science will attempt to reproduce ... orchestrations of Flora and Fauna and publish the scores" (Serres, 2010/2012, p. 133)? Ha (*he chuckles*). Along with those beautiful concertos, Henrietta, I believe your voice can be heard, and has been heard today in this place.

Lacks: So how do we make these changes? War, arguments, and lawsuits rarely bring about any gains or changes.

Serres: I agree, "Neither debate nor criticism makes any advances, except on the social chessboard and in the conquest of power" (Serres & Latour, 1990/1996, p. 37). What must happen tor "advancement in philosophy and also in science, is inventing concepts, and this invention always takes place in solitude, independence, and freedom—indeed in silence" (p. 37).

Hoffen: In solitude, maybe your voice can be heard, and the voices of others like you. Each of us must find a space where transformation can take place. It will be in solitude that we decide to tell your story and the story of others, or to remain silent.

Serres: (*to Hoffen*) You spend time in solitude, studying, very passionately. That is why you can see and hear Henrietta. I assume you are an educator. Or maybe a graduate student? What can you tell us about what science and science education should encompass?

Hoffen: "The true purpose of education was to prepare people to deal with ... socially relevant questions—to equip them for the age in which they live" (Deboer, 1991, p. xii). If we look and "see what has been lost and gained by each major shift in the past," we can "head with greater vigor in the direction of social responsibility and socially relevant instruction in science" (p. xii). As we move forward in time, there will be more stories like Henrietta's. In an age where the human genome moves to the forefront of science, there will be other ways in which people are segregated, not by race, or sex, but by specific genes. The idea is to continue to move forward. We must pay attention to the stories along the way. We must often revisit the past and hover a moment ... Answers we seek maybe in something that science has already seen.

Serres: "Time flows in an extraordinarily complex, unexpected complicated way ... it twists and folds" (Serres & Latour, 1990/1996, p. 58).

The group looks up and Henrietta has vanished into the crowd. They are all quite bewildered at the day's events.

Landecker: Where did she go? We have certainly experienced a "fold" in our day (Serres & Latour, 1990/1996, p. 58).

Rheinberger: I like that reference, Hannah. "The history of science understood as the past of a present-day science, is by no means a blueprint, read backwards, of the chronological unfolding of a given science in a given period" (Rheinberger, 2010, p. 39). It is stories, many stories that can be linked together in multiple ways.

Serres: "Philosophy can be summed up in little stories" (Serres & Latour, 1990/1996, p. 25). Have we not all experienced many small stories in this session; stories from science, stories from scientists, OncoMouse, HeLa cells, Henrietta, Henrietta's family, and stories about each of us as we participated this morning?

Rheinberger: Together we have brought what we know about science, our stories, and our cultures. "Science might even be described as the most profoundly 'historical' cultural enterprise, since scientific achievements are defined precisely by the possibility in principle to become suspended. Science is only science as a constant process of becoming" (Rheinberger, 2010, pp. 40-41).

Landecker: Another layer of complexity in this hierarchy of stories is how Henrietta's "story is simultaneously what happened to a person and her body and a narrative vehicle through which journalists and scientists have imagined and witnessed the possibilities for lives and bodies constantly being changed" by a system of ever-evolving technologies that are of human origin (Landecker, 2000, p. 54). The interaction shows us how science is made.

Latour: Amazing discourse, my colleagues ... science, HeLa, and a mysterious visit from Henrietta.

Hoffen: Dr. Rheinberger—I have an idea for your #10 on the list. How about—Henrietta Lacks. Maybe stories behind scientific developments. Hmm ... maybe that is too narrow ... How about: Bring stories to science; let Henrietta and stories of unknown others be told, shared, and created. Maybe through lingering participants can create complex connections with stories of science, participants of science ... Maybe ...

Hoffen's head is spinning with ideas. The group continues to talk to members of the audience.

POSTSCRIPT: COMPONENTS OF CURRICULUM

Creating curriculum is not a simplistic endeavor. Curriculum should be fluid and flowing or in constant motion like a pendulum or seesaw. As curriculum moves through time, touching each participant, it grows. "The most elaborate narratives, myths, and icons always return us to this point of chiasmic see-sawing" (Guattari, 1995, p. 98). During this back-and-forth process ... like a pendulum swinging, "something is absorbed—incorporated, digested—from which new lines of meaning take shape and are drawn out" (p. 96). Teachers must expose students to opportunities in the classroom that create opportunities for lingering and discourse in thinking. Rushing from topic to topic to meet a curriculum map does not support lingering or discourse. Students and teachers must have time to take in information, think about it, and then examine them for new meanings. Because, "Scientific discourses are lumpy; they contain and enact condensed contestations for meanings and practices" (p. 204). If teachers continue to give out solitary answers and give an impression to students that they are all-knowing there will be an absence of learning. Through conversations concerning stories relating to the curriculum, thought processes can be disrupted and new ideas constructed. All participants of curriculum must be able to linger, to "wander and wonder" through the world, conversations, and encounters (Schubert, 2009, pp. 231). With all the different wanderings and wonderings, the new fabric could be woven, something new that has never been before. I would think this new fabric could become a bridge between old perceptions and newly developed perceptions.

Jacques Derrida assures us that flaws do not reside in science or one's philosophy alone. He tells us "indeed, one must understand this incompetence of science, which is also the incompetence of philosophy, the closure of the epistémè" (1974, p. 149). This incompetence opens new places for new perspectives and new knowledge. Neither science nor philosophy is ideal; both must incorporate as many visions as possible. We as educators and students must explore the space in between the two entities: "The natural tendency of theory—of what unites philosophy and science in the epistémè—will push rather toward filling in the breach than toward forcing the closure" (p. 148). By generating stories, a third space between science and philosophy is created. "Stories are a core aspect of the constitution of an object of scientific knowledge" thereby giving them authority in the science classrooms in a postmodern society (Haraway, 1991, p. 82). It will be the culmination stories of scholars, educators, students, and the scientists, who all become characters in an ever-changing curriculum that will "fill the breach" in today's science classrooms (Derrida, 1974, p. 148).

CHAPTER 5

CONSPIRING TO INTERRUPT SCIENCE AND SCIENCE EDUCATION

PRESCRIPT: ENTER HENRIETTA'S IMMORTAL CELLS

Henrietta spent two days at Johns Hopkins Hospital following her first radium treatment. Down in George Gey's lab, Mary Kubicek was taking care of those famous test tubes labeled HeLa. "Mary started her days with the usual sterilization drill. She peered into the tubes, laughing to herself and thinking, *nothing's happening. Big surprise*" (Skloot, 2010, p. 41). Henrietta left the hospital with an appointment to return for another radium treatment 2 and a half weeks later. Just 2 short days after Henrietta went home to Turner Station, something odd occurred down in George Gey's lab. Mary observed "little rings of fried egg white around the clots at the bottoms of each tube" (p. 40). Henrietta's cells were surviving, "they were growing with mythological intensity" (p. 40). The cells grew and filled up "as much space as Mary gave them" (p. 40). George Gey and Mary watched with caution. Both were expecting the cells to die suddenly—like all the other cells over the last 20 years. Dr. Gey "told a few of his closest colleagues that he thought his lab might have grown the first immortal human cells. To which they replied, Can I have some? And George said yes" (p. 41).

It is obvious at this moment in time that the last thing on George Gey's mind is dollar signs or issues of who owned Henrietta's cells. Gey was in

Mandy Hoffen and a Conspiracy to Resurrect Life and Social Justice in Science Curriculum With Henrietta Lacks: A Play, pp. 105–138
Copyright © 2021 by Information Age Publishing

this work for the science. Originally, Gey wanted to find a cure for cervical cancer. Henrietta died of that disease in October 1951. Shortly thereafter, planning began for a HeLa factory, a massive operation that would grow to produce trillions of HeLa cells each week. It was built for one reason: to help stop polio "in the midst of the biggest polio epidemic in history. Schools closed, parents panicked, and the public grew desperate for a vaccine" (Skloot, 2010, p. 93). Jonas Salk had "developed the world's first polio vaccine, but he couldn't begin offering it to children until he'd tested it large scale to prove it was safe and effective" (p. 92).

The National Foundation for Infantile Paralysis had been using monkeys to test Salk's vaccine. All the monkeys were "killed in the process" of testing (Skloot, 2010, p. 94). Today this would have been a huge issue of animal welfare, but in 1952 the problem was one of economics … monkeys were expensive. Testing Salk's vaccine on monkey cells would cost millions. The National Foundation for Infantile Paralysis contacted George Gey. This organization was offering cell culturists a possible $50 million to mass produce cells for research. George Gey began working out the details.

In April 1952, approximately 1 year after Henrietta's cells had become the first immortal human cell line, George Gey and William Scherer, both now serving on the National Foundation for Infantile Paralysis advisory committee, tried infecting HeLa cells with polio virus. The surprised researchers were amazed that HeLa cells could easily be infected with polio virus. The two very excited researchers began experimenting with ways to ship HeLa cells all over the world. Scherer was chosen to oversee operations of the new HeLa cell distribution center to be located at the Tuskegee Institute. Charles Bynum was the "director of Negro Activities" for National Foundation for Infantile Paralysis (Skloot, 2010, p. 96). Bynum, a science teacher and civil rights activist "was the first Black foundation executive in the country" (p. 96). "He wanted the center to be located at Tuskegee because it would provide hundreds of thousands of dollars in funding, many jobs, and training opportunities for young Black scientists" (p. 96). *The New York Times* (1955, p. 25) published a story with a headline reading: "UNIT AT TUSKEGEE HELPS POLIO FIGHT: Corps of Negro Scientists Has Key Role in Evaluating Dr. Salk's Vaccine." The news story included photographs of African American women examining cells with microscopes. "Black scientists and technicians, many of them women, used cells from a Black woman to help save the lives of millions of Americans, most of them White" (1955, p. 25).

This complex chain of events involving taking a woman's cells, growing them in a laboratory, using them to test polio vaccine, creating a HeLa factory, and opening up opportunities for African-American scientists to work in the field of scientific research, invites us to see that *The Immortal*

Life of Henrietta Lacks is not just the story of an ordinary Baltimore house-wife who died of cervical cancer. Her cells are alive today. The story of her cells not only chronicles success in defeating polio, but the cells created a pathway for women and persons of color to enter the field of science. What began as a simple act of collaboration and sharing of cells had now turned into a huge operation. We know what this operation meant for science. The eradication of polio means that every man, woman, and child on the planet has benefited from Henrietta's cells. We have immunity to polio pulsing through our bodies all because scientists were able to test the polio vaccine on Henrietta's HeLa cells. The story does not end here, it provokes many questions. How would Henrietta have felt about her cells being used? What happens to Henrietta's family? What aspects of this story need to be part of science curriculum? What parts of this story should be part of the conversation concerning science and science education?

In a world where traditional ideas of science seem prevalent, John Weaver, Peter Applebaum, David Blades, Patti Lather, and Nancy Cartwright each bring a unique perspective on why science education must change in order to meet the needs of a changing society. Through a careful examination of science, illuminated by curriculum theory, these authors bring unique perspectives that could make science a vehicle for creating a more democratic society.

ACT V

In Act V, I hope to unravel this idea of what science is in reference to today's society, and all its cultures. Economics can impact both society and cultures. Can economics impact the doing of science? Did the discovery of HeLa cells start the precedence for scientists to consider science beyond how it affects participants and society? How is science impacted by the commodification of the human body, human organs, or human cells like HeLa?

Act V: Conspiracy to Interrupt Science and Science Education

Characters in order of appearance: John Weaver*, professor; Mandy Hoffen, graduate student/teacher; Peter Applebaum*, professor; Nancy Cartwright*, professor; David Blades*, professor; Patti Lather*, professor; Henrietta Lacks; Security Officer; Polly Merase, Cellular News Network reporter; Priscilla Wald*, professor; Hannah Landecker*, professor.

Members of the panel make their way to the front of the room as audience members file in and take their seats. The panel moves the chairs from behind the table and form a semi-circle near the audience, filled with mostly graduate students, and a few university professors. After a short time of greetings and hand shaking, John Weaver checks his watch, opens his notebook, and addresses the panel.

> **Weaver**: It is 11:00. I say we get started. Maybe we should start with introductions for our audience. I am John Weaver. I had a very interesting journey to Johns Hopkins. You might say I got lost. (*He smiles at Patti Lather as he says this. Patti Lather [2007] had published a book called* Getting Lost.)

> **Lather:** So John, you are saying that you found yourself "in an awkward position that was not so much about losing oneself in knowledge as about knowledge that loses itself in the necessary blind spots of understanding" (Lather, 2007, p. vii)? We must take time to examine those blind spots in our thinking. What did you find while you were getting lost?

> **Weaver:** First, Patti, I found myself physically lost. During this excursion, I had the opportunity to learn much about the many obscurities associated with the story of Henrietta Lacks. I managed to find myself in Clover, Virginia. Any of you ever been there? (*He looks out at the audience and shakes his head slightly.*) Henrietta Lacks was born and buried in Clover. She was living in Turner Station when she was a patient here at Johns Hopkins. It is Henrietta's cells that this conference is cell-e-brating ... did you get my play on words? I was dumbfounded to have accidentally found myself lost in Henrietta's hometown. It was all somewhat surreal, considering the focus of our conference. As I was looking for the jack in the trunk of my rental car to change a busted tire, a man approached me. His name just happened to be Cootie Lacks. Cootie is Henrietta's cousin. There was no tire jack. Fortunately, for me, Cootie was a very animated storyteller. He shared Henrietta's story with me from a very different perspective. Henrietta's story, until Rebecca Skloot's 2010 book, *The Immortal Life of Henrietta Lacks*, has rarely been told from the point of view of her family. This area might be what Patti would call a blind spot. And yes, I had to be willing to listen, as Cootie poignantly pointed out to me. I was lost, had a flat tire, and no cell phone service. I was getting nowhere fast. Hearing Cootie's perspective of this

story allowed me to develop a new perspective on Henrietta's story.

My goal here today is to make sure we examine science from many perspectives. I want us to linger and consider these topics deeply. I am very interested in taking on the task of securing a science education for students of all ages, all cultures, and all backgrounds that is applicable to the world in which we live. I think I can say that collectively all the members of our panel today are very "interested in banishing what Bruno Latour in his book, *Pandora's Hope*, calls a brain-in-a-vat approach to science" (Latour, 1999, p. 4). Think about a brain disconnected from the body. Everything that can possibly simulate this brain must be provided from the outside. The brain-in-a-vat "approach exempts science from being responsible for that which it creates" (Weaver & Anijar, 2001, p. 247). The panel assembled here today is looking to find a replacement for this brain-in-a-vat approach. A science that is connected to the world.

We feel that science can be resurrected with "pedagogies of the cultural studies of science" (p. 247). Donna Haraway says we want students who are "bumptious technoscience actor[s]" (Haraway, 1997, p. 94). By using the word bumptious … she is stressing that students must take an enthusiastic passionate role as participants in their world of science and technology. Today, I want to make connections that will create a new perspective for science and the students of science where "science is not dismissed nor worshiped but utilized to construct a democratic and sustainable world" (p. 247).

Appelbaum: I kind of like that word … bumptious … Can students be bumptious learners in a standardized environment with all the current standardized testing? Hmm. (*He lingers deep into thought.*) Oh, sorry, I am Peter Appelbaum. I think we must find a way to have real conversations "about the pedagogies of science" at all levels of education, secondary education included (Appelbaum, 2001, p. 111). Philosophers and historians of science often "under theorize education and inadequately address educational institutions of science" (p. 111). Meanwhile, those participating in these ongoing debates "are curiously absent from contemporary educational studies. And educators often accept a stereotyped and monolithic perspective on science" (p. 111). There is a disconnect between the theorists working at the university level and teachers working in public schools. There are many connections to science that we must explore. I would be interested to see "how might a conversation among scientists, and colleagues who teach science methods and those in curriculum theory unfold" (p. 124). I use the term science methods loosely here. I am not speaking only of the old-school scientific method. I am referring to professors on the university level who are teaching their students how to teach science.

Cartwright: I know it is not my turn, but I must say Peter that I think physicists might be guilty of undertheorizing the applications of physics to real-world situations. "My ultimate concern in studying science is with the day-to-day world where SQUIDS (superconducting quantum interference devices) can be used to detect stroke victims and where life expectancy is calculated to vary by 35 years from one country to another" (Cartwright, 1999, p. 5). Yes, you must know your physics ... but how the physics is applied is what should be stressed. "I look at the claims of science, at the possible effects of science as a body of knowledge, in order to see what we can achieve with this knowledge" (p. 5). I approach this issue with "the motive of a social engineer" (p. 5). There is more to physics than correctly calculating the appropriate mathematical formula. I think we need to ask a bigger question. "How can our world be changed by science to make it the way it should be?" (Cartwright, 2015). Peter, I like how you are using the term method.... "The scientific method is a broad church indeed," a church that should be opened up to encompass more than a strict plan that is thought to be applicable to all situations (Cartwright, 2015).

Weaver: I think we might have to call Nancy here bumptious. Science for the sake of science is not very meaningful. I agree, the meaning is in those real-life connections. We could probably spend a whole day talking about scientific method. Mandy Hoffen, my graduate student sitting here on the front row, wrote a paper about the scientific method taught in schools called "Pink Rats Have Excellent Racing Cars," early in her graduate program. She employed this mnemonic device to help her students remember the steps of the scientific method: problem, research, hypothesis, experiment, results, and conclusion in the right order for their standardized test. I know from Mandy's paper that she discusses method in broad terms with her students as well. However, the students must be able to regurgitate this specific method. I hope today that we compose a plan to make sure that the steps are not all that students are learning in science class. Shall we continue? (*He motions to David to continue the introductions.*)

Blades: Good morning, I am David Blades. I am here to speak on behalf of students of science. Learning scientific method is not a horrible thing. I like the use of the mnemonic device, Mandy. It is clever. I would venture to say that you let your students create that mnemonic. Empirical science has a place … but there has got to be more.

> After years of learning to regurgitate information that has no relevance to their lives and no particular significance, except for the few who become professionally involved in the discourse of modern science, their vision and curiosity die, leaving only a lingering disgust that after so many years of learning science they really have learned little that is worthwhile. (Blades, 2001, p. 72)

My heart's desire is to return science to a place where we, students and teachers, "find hope" (p. 82). Hope for a curriculum that is meaningful.

Cartwright: Now that it is my turn, my name is Nancy Cartwright. I am a physicist. I was thinking that I need to apologize for going out of turn … but then I began thinking that being called bumptious is an extreme compliment … so I will not apologize. I must say that I am feeling a little conspicuous about sitting among such a distinguished group of curriculum theorists. Being called bumptious is new but feeling conspicuous in a group of scholars is not. I must confess, "I differ so radically despite our shared interests" from leading philosophers of physics (Cartwright, 1999, p. 5). Most of "their interest in science generally comes from their belief that understanding our most advanced scientific representations of the world is their best route to understanding that world itself" (p. 5). I am interested in science from a different perspective. A perspective concerned with, "not primarily the little politics of laboratory life" but life "that shapes the internal details of science," details such as "big politics that builds bombs and human genomes" (p. 5). My perspective and interest in science is "different again from most historians and sociologists of science whose immediate object is science itself, but—unlike philosophers who use our best science as a window to the world—science as it is practiced as a historical process" (p. 5).

Weaver: Nancy, you are in good company; everyone on this panel has written books that are different from most books written about education. Proceed!

Cartwright: The name of my most recent book is *The Dappled World*. "The laws that describe this [dappled] world are a patchwork, not a pyramid. They do not take after the simple elegant and abstract structure of a system of axioms and theorems. Rather they look like—and steadfastly stick to looking like—science as we know it: apportioned into disciplines, apparently arbitrarily grown up" (Cartwright, 1999, p. 1). "The dappled world is what comes naturally; regimented behavior that results from good engineering" (p. 1). I work to contest the laws of science that seem to be carved in stone. I have a question to ask here. Does science curriculum and science education continue to address a science that is not representative of all that science encompasses? David, I worry about my students as well. I think students are well versed in some aspects of science. They know, as do I, that an oak tree will grow from an acorn, "but not from a pinecone; that nurturing will make [a] child more secure; that feeding the hungry and housing the homeless will make for less misery; and that giving more smear tests will lessen the incidence of cervical cancer" (p. 23). I am a physicist, but as you can see all these topics relate to me as a human being.

Lather: I have to confess Nancy, I had to Google SQUID. I have never heard about superconducting quantum interference devices. It sounds like something out of a science fiction movie. My first thoughts were of the ink-shooting cephalopod.

Appelbaum: That was at least a scientific stab in the dark, Patti. My brain went right to calamari, fried with some type of spicy sauce. It must be getting close to lunch time. Please forgive my digression and continue.

Lather: It is interesting that someone might be able to have an EEG (electroencephalogram) without all those electrodes attached to their bodies. The sensitivity of the technology is amazing. I will send everyone the link I found about SQUID devices. I am obviously not a physicist, but I do have a love for learning new things, especially when they relate to the world, I live in. Oh, my name ... I am Patti Lather. I want to talk about detours for a moment. John may have had the most interesting detour, meeting Henrietta's cousin, Cootie ... but his story describes an aspect of learning that needs to be part of science and science education. I have studied and written about this phenomenon of getting lost for several years. What would happen if those of us given the task of writing curriculum and educational policy could spend some time "getting lost" and really taking in all that educational policy includes? "Perhaps 'getting lost' in policy work might exactly be about accountability to complexity and the political value of not being so sure" (Lather, 2010, p. 16). Policymakers might begin to see their work from a different point of view.

Patti continues speaking but her attention is drawn to the back of the room. No one else seems to notice the woman standing on the very back row, in the shadows. The shadows make it very difficult to see the woman's face. It seems that Patti is the only one in the room who can hear this woman speaking. Wait a minute, she thinks to herself—this is the woman she saw downstairs, who asked about the public ward. Patti strains her eyes to see the woman and strains her ears to hear what the woman is saying, although no one else seems to be able to hear her.

Lacks: My name is Henrietta Lacks. I was known as only HeLa for almost 25 years after my death. The doctor took MY cells without MY permission. The doctors did not even tell my family that my cells were alive. Everyone thinks it is such a wonderful thing that my cells went to the moon ... helped cure polio. Did anyone ever think about how my family would be impacted by my death, and then the news of my cells? Everyone thinks of me as another Baltimore housewife who died of cancer. What happened to me was wrong! What happened to my family was horrible ... They scientists did nothing to help my family. (*She is getting louder.*)

Lather: Now calm down, there is no need to yell.

Cartwright and Blades look puzzled when Lather makes this statement.

Lacks: No reason to yell. Did my own death not silence me? Science took my voice away! Was I not silent for over 25 years while my family wondered what happened to me here ... here where the night doctors would snatch people out of the streets for their "RESEARCH" (Skloot, 2010, p. 165)? Those writers calling me a "monster among the Pyrex" (Landecker, 2007, p. 171). I will show them a monster. (*She leaves her seat, looking around at the audience members.*)

Mandy turns around and sees the woman standing, now charging toward the aisle. A security guard enters from a side door and makes his way to the stage and whispers to Weaver.

Guard: Keep an eye out, we are trying to find a woman in a brown suit, with a fancy collar. She is approximately five feet tall. We have film footage of her leaving the exhibit hall with the stolen suit. We have no film footage of her entering the hallway. We need to question her about taking the clothes from the mannequin. Keep your phone tuned in to CNN.

Weaver: (*puzzled*) CNN? Conspiracy to interrupt my session News Network.

Guard: The Cellular News Network. Here at Johns Hopkins we have our own social media for all the scientists. The patients even log on and leave DNA sequences. (*He sees that Weaver is puzzled again.*) You know notes, messages, and tweets. This is a science institution. We call them sequences because …

Weaver: Yeah, yeah, yeah, I get that. They teach about DNA where I come from too. This conference has been so full of interruptions. Let me get my session back on track. We will keep an eye out.

Lacks: Why did science not want anyone to know the cells came from me? Did my race keep me from getting treatment that could have saved my life? Why did the doctor take my cells without asking? Why are my children without health care? Why are corporations allowed to keep padding their pockets with profits from the sale of my cells?

Lather: Sale of your cells … that sounds funny! Sorry, I am not making fun of you. I just liked how the words sounded. Stay calm. I know you are upset. I read about it all this morning.

Henrietta temporarily fades into the shadows.

Weaver: Thanks Patti, I also want others—university professors, teachers, policymakers, government—to have a broader view of science and science education. We need to look at science through a lens that "embraces the ambiguous, uncertain, and infinite" (Weaver, 2001, p. 2). It is time for a science that refuses to neither adhere to traditional method, be all knowing and powerful, nor be unreachable by ordinary people. By teaching students, a strict scientific method—Problem,

Research, Hypothesis, Experiment, Results, and Conclusion—students get the impression that there is only one way to solve a problem. Oh, I almost forgot. The guard that just interrupted says we all (*he looks at the audience and gestures that they are part of all*) need to pull up CNN on our phones or other devices ... Security is looking for a woman. There is a picture posted on the base page ... can everyone see it? She is dressed in a brown suit that oddly resembles the suit taken from the mannequin in the exhibit hall. For those of you that do not know, there was an art exhibit destroyed last night when a neon sign shorted out. You all now have the link. We must get back to our session.

Appelbaum: John, I want to interrupt here a minute. There is a press conference going on about the incident in the exhibit hall last night. I think I can pop it up on the screen for everyone to see and hear.

Weaver: We may as well take a moment and watch—this might help everyone get on the same page. Okay Peter—link us up!

The projector comes on, with the live news report.

Reporter: This is Polly Merase reporting live for Cellular News Network at The World Alliance of Science and Science Education Symposium. We are following a story in Baltimore at Johns Hopkins Hospital and Research Center. Last evening during a thunderstorm, an electrical outage occurred, and this BioArt display was destroyed. (*Cameras show a close-up of the exhibit, complete with the neon pink "Henrietta's" sign, a microscope, a beaker, a notebook, and a picture of Henrietta in a brown suit.*) This rather unusual exhibit consists of a bright pink neon sign and a working microscope. People viewing the artwork can see HeLa cells magnified 1000 times and read about the original cell donor, Henrietta Lacks. Investigators believe an electrical short from this sign caused the cells in this beaker to be spilt all over the floor. When forensic scientists arrived on the scene early this morning, they reported that the original HeLa cell culture on display had disappeared. To the left of the exhibit, as you can see, we are left with an unclothed mannequin. A brown suit, vintage 1951, and brown leather sandals were removed from the mannequin. Although the investigators have positive tests for culture media, there are no positive cultures for HeLa cells. The artist reports the spilt broken beaker should have been overflowing with HeLa cells. Currently thousands of educators, professors, and scientists are gathered to celebrate HeLa cells. A woman wearing the suit and shoes was seen leaving the exhibit hall this morning around 7:45 A.M. We believe this woman may have answers to our questions. Please text CNN, Cellular News Network, #HeLa, along with the information, if you see this mysterious woman, or have any information concerning this mysterious event. Meanwhile the artist, Pierre Phillipe Freymond (2006), has refurbished the exhibit, and the exhibit hall is opened. The mannequin has been dressed in a new suit and new shoes provided by the History of Fashion museum and its curator, world-renowned fashion expert Dr. Mimi Pepys Farrell. We will continue to cover this story.

Lather: John, while we are amidst interruptions, are you going to address the questions raised by the woman in the back?

Weaver: Did someone have a question? (*He shrugs his shoulder and continues when he sees Patricia Wald come into the lecture hall.*) Let's all welcome Patricia Wald. (*Patricia comes down the aisle. She tells Henrietta Lacks "excuse me," as she walks by her.*)

Wald: I am so sorry to be running late. There is chaos downstairs ... You would think some conspiracy is afoot to ruin our carefully planned conference. Missing clothes ... now missing HeLa cells. CNN is getting reports that this mysterious woman in the brown suit attended a conference session this morning, and even spoke to a group of philosophers ... Honestly, I don't know if the philosophers were speaking metaphorically about Henrietta Lacks or if they actually spoke to her.

Mandy slides down in her seat slightly. She is not sure if she had seen Henrietta either. Looking around the room nervously, she was still trying to figure out if the whole conversation that took place with Henrietta was real ... or if it just happened in her head.

Wald: Let's quickly get to our topic or topics for the day. I am very happy to be here with all of you ... we have important work to do concerning science and science education. I presume that I have only missed your introductions. And who will be taking notes to submit to the conference officials?

Blades: I will take notes. (*The others smiled, happy not to have the responsibility.*) Now you want a list of what we think is needed to improve science and science curriculum in schools, right?

Wald: I thank you, David. That is correct. Now we can move on with our discussion, without interruptions, hopefully (*she sounds unsure*). Wait, I never introduced myself. I am Patricia Wald. I work in the Department of Literature at Duke University. In 2008 an essay was published by a multidisciplinary group from Stanford University. This group outlined "principles designed to serve as guidelines in the debates that have surfaced with the acceleration of research in human genetic variation following the mapping of the human genome" (Wald, 2012b, p. 248). In their meeting, "They urge medical researchers to minimize their use of the categories of race and ethnicity and to be especially careful about how they explain those labels" (p. 249). I start here today because I believe that the story of Henrietta Lacks "offers important insight into the deeper issues of institutional racism that are frequently overlooked" (p. 249). This story "calls attention to social inequities that are at the center of debates surrounding health care, but it is equally relevant for the insight it can offer into the spirited debates surrounding race and DNA at present" (p. 249). Let's talk about HeLa cells first. "The HeLa cell line represented a radical breakthrough, but it also created a new organic entity— a Frankenstein's creature—for which there was no context" (p. 249).

Lacks: (*Shouting.*) Frankenstein's creature … Pyrex monster (Landecker, 2007) … (*She gets up out of her seat and starts toward the front of the lecture hall.*)

Wald: George Gey had been trying for years to grow cells outside the body. Suddenly, he had more than he knew what to do with. The creation of this cell line quickly began "exemplifying some of the ways in which scientific research was challenging conventional biological definitions of the term human being" (Wald, 2012b, p. 249). "Many involved in the HeLa research noted its resemblance to science fiction. Indeed, that genre crystallized around these questions in the post war moment and offers theoretical insight into the instability of the concept of human being" (p. 249). We have to ask … what does it mean to be human?

Lacks: Science fiction … do you think what I lived through was science fiction? (*She takes the microphone out of Patricia Wald's hand.*) I was just a mother, a housewife, in Baltimore, trying to raise my children. My youngest daughter, Deborah, had nightmares about me being blown up in the space shuttle. Would I be alive if I had not been treated in the public ward? (*Lacks's voice gets remarkably quieter. She looks at her hands.*) I died in that public ward. All this about immortal cells. I died. I died a very painful death. Now I am "Alive! Still alive. Alive again" (Butler, 1987, p. 3). How is it that I am here now walking the halls of Johns Hopkins, what else has medicine hidden from me? From my family?

Henrietta drops the microphone and looks fearfully around the auditorium. Mandy is still unable to believe her eyes. Did she speak with this woman during the last session? Is this woman Henrietta Lacks? What will law enforcement do to her if they find her?

Hoffen: (*Mandy reaches out and touches the hem of Henrietta's jacket and whispers.*) I know your cells have saved millions of people … you deserve to be here. You deserve the opportunity to speak … but Henrietta, you must go. They are looking for you. You must go hide. I will come find you later. Find a safe place.

Lacks: I have not done anything wrong. Why are they looking for me? My cells saved humanity from polio … from cancer. Aren't they here to honor my cells? Why can't I be a part of this? I thought it was a celebration for me.

Hoffen: You are a part of this, you made all this possible. They have you confused with someone who destroyed an art exhibit last night.

Henrietta backs away and begins walking up the aisle. Again, no one seems to notice her. Wald thinks she has dropped the microphone; she reaches down and picks it up and continues speaking.

Hoffen: (*Talking quietly to herself.*) Did anyone else see what I just saw? Has Henrietta Lacks been resurrected from the grave? But how? A vat of HeLa cells? From a BioArt exhibit? (*Maybe her imagination was taking over. Her head was spinning. Wald continues to speak, unaware of the previous disturbance.*)

Wald: As I studied HeLa cells and how they were being used in laboratories, and studied Henrietta's story, I had to ask, what exactly does it mean to be human in this scientific and technological age?

Security guards and local police burst into both sets of double doors of the lecture hall.

Hoffen: (*Again, muttering to herself.*) Oh no, someone told them she was here. That means someone else saw her too. Or maybe this is all a conspiracy to drive me crazy.

Guard: Where is she? We got a sequence on CNN reporting that a woman in a brown suit entered this room. (*The guard gets many puzzled looks. The panel is becoming annoyed with the interruptions.*)

Lather: I saw the woman. She hurried out a few moments ago. I saw her downstairs this morning. She arrived when I did this morning. She asked the receptionist for directions to the public ward.

Hoffen: You mean you saw her too? Did you see her come up on the stage and take the microphone? She was very angry.

Wald: Whoa, we must continue our session. All this craziness … Took the microphone from my hand? I just dropped it. (*She looks at Mandy on the front row.*) Are you sure you are not seeing things? John told me earlier today that your research has elements of storytelling and science fiction … Could this all be happening in your imagination?

Patti Lather looks at Mandy sympathetically. She was questioning her own sanity at the moment. If she had not been googling Henrietta on her tablet she might have seen more and confirmed the graduate student's account of Henrietta coming up to the stage and grabbing the microphone. Were two parallel universes intersecting today?

Lather: (*Cheerfully.*) Let us get on with our session. We will let ZNN or CNN and the local police worry about this woman walking or running around … looking like Henrietta Lacks. I am sure it is just a case of mistaken identity.

Wald: Now, back to my question: What exactly does it mean to be human in this scientific and technological age? "The creation of new and unfamiliar organic entities, such as cell lines, commingled with the haunting images of human beings stripped of their humanity to challenge in their uncanniness conventional definitions of human being and humanity" (Wald, 2012b, p. 249).

Lather: Images seem to make an imprint in our mind. Those images sometimes do haunt us. Maybe we are all seeing Henrietta's ghost. She did die right here at Johns Hopkins Hospital. Personally, I think the terms haunting or ghost could have negative connotations concerning Henrietta ... I think that takes us back to Gartler's racialized language that Hannah Landecker mentioned this morning. I think we need to return humanity to Henrietta's story. View her with a different lens. What about angel? Maybe we need to think of Henrietta as an angel—"a necessary angel" (Lather & Smithies, 1997, p. 194). "Necessary angels are about our need to create and believe in what we can hardly avoid suspecting are fictions" (p. 174). Necessary angels become witnesses, "to the human capacity to carry on and even sing in the midst of anguish, an audience of astonished angels, come down to earth to learn about living in an historical time of permanent emergency" (p. 174).

Wald: Ghost, angel ... all of this unfolds like a play being performed on stage. The plot is very complex. Science has a reputation of operating on facts generated by experimental research. In so many cases the human side of the story is not cast. By having conferences such as this one, where we allow the human side of the story to enter the conversation, "The focus on the human drama is significant for its reminder of the importance of the human dimension of research in medicine, science, and biotechnology" (Wald, 2012a, p. 249). Humans in your classrooms are stakeholders. Not only in their learning, but in how they will participate in real-life, real-world science.

Blades: I think this is a good place for me to interject. The students are indeed stakeholders. Stakeholders on the bottom of the power block. I want you to think about science class from the students' point of view. Think about a student's journey as a story. Typically, it all begins when a young student may be asked to learn the parts of the flower using proper terminology—stamen, pistil, sepals, style, stigma, to name a few. "By the middle years of their education they can list from memory, at least before the test, the scientific plant groups" (p. 72). In high school, "students are invested in the taxonomic schemes of botany and the biosynthetic details of photosynthesis, information they had better regurgitate for the final exam if they are to obtain the grades necessary to enter the university" (p. 72). The child does not "touch the plants," or "marvel at the struggle of a dandelion pushing through the concrete or study the ancient use of herbs: That is not science" (p. 72). Finally, the students make it to their graduation, suffering "the terminal stages of bulimia" (p. 72). Although it may not be the most pleasant word, I think it does describe how horrible current practices in science have become from the students' point of view. "Teachers naively begin the slowly destructive bulimia of school science education by emphasizing to the children how the enterprise of science deals in facts" (p. 72). Learning should be fun ... Learning should never have to be equated with a dreadful eating disorder. However, "the bulimia of science education reaches advanced stages as student's classroom experiences of science increasingly feature nice, tidy, banal experiments that have little to do with reality" (pp. 74–75). Personally, I have always been "impressed with the ability of stories to capture the attention of my students" (Blades, 1997, p. 7). I was quite taken with the story I heard from Rebecca, Hannah, and Patricia this morning. I understand Mandy's students have been quite taken with the story of Henrietta Lacks as well.

Mandy, on the front row, talking to herself under her breath, did not hear Dr. Blades mention her name.

Hoffen: She seemed pretty real to me. Let's see if I can make sense of all this craziness. So, it began when HeLa cells took over the lab, years ago; Henrietta was not even mentioned. Now Henrietta is back ... taking over this conference. But how? How did she get here? Maybe she never left.

Weaver interjects before Blades can repeat his comment.

Weaver: Bulimia is an excellent way to describe the process, David. Students should be discussing "how scientists construct models to understand the world, and these models may produce evidence to support their usage, but they are only models or simulations, not correspondent representations" (Weaver, 2001, p. 15). With bulimia, they just spit back an interpretation of a model their teacher has explained. The only interpretation of the model they experience is the one the teacher gives them.

Appelbaum: All the emphasis on standardized testing has changed the role of the teacher. It has been my experience to find "students in science classes enhancing their propensity to interpret the world as a scientist. Instead, they spend most of their time learning to parrot already-developed techniques and applications of science" (Appelbaum, 2001, p. 114). Students should have to read, write, and think about material that they need to learn. Giving them the answers defeats that purpose. Force feeding students large amounts of information in worksheet packets to make them proficient standardized test takers is not something most students will find appealing. "Students learn best when they care about what they are learning, when what they do matters, when there is a purpose, a commitment to what they are doing" (p. 117). Oddly enough, it must be their idea to become a part of what is offered in a classroom. If we are fortunate enough to coerce the student into reaching out and wanting to learn new information and new skills, something wonderful occurs. A magical connection is created between students and their learning. Think of it as a bridge. However, the students must build this bridge themselves. The teacher should be cast in the role of facilitator:

> Teachers must facilitate interaction at the frontier where information of the science disciplines intersects with understandings and experiences that individuals carry with them to school; they help students to interpret their own lives and encounter new propensities as a result of their encounter with school. (Appelbaum, 2001, p. 115).

Blades: "Forming and sharing narratives can become a call to action through a commitment to change present situations" (Blades, 1997, p. 8). This call of action involves "the individual sharing the story" and "those listening or reading the stories" (p. 8). The process directs participants to rewrite their own stories based on what they are hearing or reading. "What can emerge from such sharing of story is a conversation of critique about our normal, comfortable, and natural participation in the world" (p. 8).

Weaver: I know we each have our own stories, but do we all have an equal opportunity to write our stories? Think about this for a moment. I think we must address that not everyone has equal opportunity when it comes to sharing, reading, and composing stories. Have you considered what might control what stories are written, told, and shared? When we talk about what it means to be human in science, we must take a giant step back, all the way back to an age where humans were bought and sold. From this perspective, I am not sure we have gained any ground. HeLa cells were originally shipped to laboratories for "ten dollars plus Air Express fees" (Skloot, 2010, p. 97). In *Rolling Stone* magazine, Michael Rogers (1976) reported the cost of a "tiny glass vial of HeLa cells, to be about $25.00" (p. 34). Henrietta's sons saw the *Rolling Stone* article. They were sure their mother's cells had been stolen and someone owed them a lot of money. Commodification of human cells ... human beings. Think about that case out in California involving John Moore and his spleen. "Dr. Gold failed to inform Moore of his intent to harvest the cancerous spleen and create immortal cell lines for further research" and probably make millions in the process. "Gold used Moore as a research object similar to the way federal scientists used African Americans in its study of syphilis," and the way George Gey and many, many others used the cells of Henrietta Lacks (Weaver, 2001, p. 21). Patricia, do you use the word bioslavery when you write about Moore?

Wald: Yes, John, I do. The questions raised in court in the Moore case also "haunted the Lacks case" (Wald, 2012a, p. 200).

Lather: Here we go again ... using haunting and ghost like vernacular, the way teachers use battle vocabulary to discuss their jobs. Sorry, I just had to interject. Why not use a more positive term for Henrietta. We could call her a guardian angel. She is still with us. She saved so many.

Wald: The term is just stuck in my head, Patti. "Issues of who owns cells, tissues, and organs became center stage in the press when John Moore sued UCLA and the courts had to make legal sense of a cell line" (Wald, 2012a, p. 200). Let me read a little bit from a paper I wrote concerning the court case. (*She removes a stapled article from her briefcase and flips over several pages before she begins reading.*)

> While the Los Angeles court ruled that Moore's cells (his discarded spleen) did not constitute personal property, and therefore that there was no case, the California court of appeal overturned the ruling on the grounds that "the essence of a property interest—the ultimate right of control—exists with regard to one's own human body," although the majority opinion conceded the need to approach the issue "with caution," since "the evolution of civilization from slavery to freedom, from regarding people as chattels to recognition of the individual dignity of each person, necessitates prudence in attributing

the qualities of property to human tissue" (Moore vs. Regents of University of California as cited by Wald, 2012a, p. 200).

Then, John, comes the term "bioslavery" (Wald, 2012a, p. 200). "The term resonated in the courtroom and the press; bioethicist and legal theorists invoked it to name the danger in the Moore case as well as for biotechnology generally (p. 200). The Moore case marked the "transformation of a constellation of issues and events into a cultural narrative" (p. 200).

Appelbaum: How humans are treated by their doctors and medical science, and related issues should become part of the cultural narrative. This is an example of science and real-life intersecting. "The role of culture in science pedagogy and the relationships among popular culture, everyday life, and school science should all be represented in science curriculum" (Appelbaum, 2001, p. 123).

Weaver: Skloot's book, *The Immortal Life of Henrietta Lacks*, also brings these issues of ownership, and the right to one's own body, to a cultural narrative. "Why has the history of science, like other dimensions of American history, become a tale of corporate gain? While the courts ruled that no individual could own their bodies, corporations could profit from these bodies without giving any compensation to the donor" (Weaver, 2001, p. 21).

Lather: Once commodification takes place, the people behind the science are forgotten. When I hear people talk about Henrietta, I think about the women with HIV that I wrote *Troubling the Angels* with ... their "whole life gets reduced to that one thing" (Lather & Smithies, 1997, p. 9). For Henrietta, her life was reduced to HeLa cells. Her life as a real live woman was replaced with the story of the cells. "We should be uncomfortable with these issues of telling other people's stories" (p. 9). We have to get it right because they are not here to tell their version.

Appelbaum: That is a sad thing to do to students, to let commodification rob them of details that humanize science. It ignores all their valuable gifts they bring to the table/classroom. Standardized testing evaluates students "at the lowest level of human thinking—the ability to memorize and mimic behavior" (Appelbaum, 2001, p. 115).

Weaver: I want you all to think about what drives the scientists to do science. Science is kind of like teaching—monetary compensation is not the motivator. Now as far as trying to get funding for research, "monetary compensation" in the form of research grants "may play a role in the decisions scientists make regarding what type of research they do, it is not the sole or major reason innovative science is done" (Weaver, 2010, p. 22). I think that "the drive to invent and discover far outweighs any monetary desire governing the actions of scientists. Even if there were little to no monetary compensation for the scientists, much of the innovative work would still be done" (p. 22).

Researchers rely on volunteers to be research subjects. "Therefore, compensating individuals for their sacrifice and involvement in a scientific or medical experiment would not end scientific invocation" unless the "entrepreneur only has an interest in profits from funneling money to scientific research" (p. 22). Compensation plans and the cost associated with such plans "will not stop the scientists from thinking and doing innovative work" (p. 22). Any compensation must be handled carefully. "I am not advocating a research model in which donors are compensated for stem cells, organs, or other body parts" (p. 22). We all know that if we start paying people for their cells, tissues, and organs, then we will have kids selling their organs to pay for their college.

Appelbaum: Did you hear about that kid in China who sold his kidney so he could get an iPad (Bennet-Smith, 2012)?

Lather: The news is full of stories like that. I was reading earlier today in Skloot's book that Zakariyya, Henrietta's youngest son, "realized he could become a research subject in exchange for a little money, a few meals, sometimes even a bed to sleep on" (Skloot, 2010, p. 208). When he needed something, he went over to Hopkins to see what was available. He allowed himself to be infected with malaria "when he needed to buy eyeglasses" (p. 208). I think Zakariyya's most profitable research volunteer venture involved a study concerning alcoholism. He used this money "to pay for a new job-training program" (p. 208). Once he "signed up for an AIDS study that would have let him sleep in a bed for nearly a week. He quit when the researchers started talking about injections, because he thought they'd infect him with AIDS" (p. 208).

Weaver: Henrietta did so much for science and her family cannot even afford their own medical care. This is tragic. What I want to propose is "an alternative to the monetary view of innovation ... when someone donates a part of their—or a loved one's—body" (Weaver, 2010, p. 22). Why not have the research university, hospital, or corporation "provide health care coverage to the donor and family for life" (p. 22). A practice such as this "would encourage people to support innovative research and help a struggling democracy overcome a health care crisis of accessibility and affordability. As it stands now, only corporations benefit from the current definition of innovation" (p. 22).

Lather: Henrietta's family would have benefited greatly from such a practice. I found it very sad in Skloot's book, when Deborah talked about the family not being able to afford health care.

Blades: John Moore would have benefited as well. The ethical issues in each of these situations are very complex. Once a consent form is signed, all the responsibility reverts to the patient. That is, if a consent form is even part of the process.

Weaver: Priscilla, I want to go back to your mention of Henrietta as a Frankenstein creature.

(Mandy cringes on the front row. This is what had upset Henrietta earlier. Where did she go?)

Weaver: As Priscilla said earlier, with the creation of a cell line from Henrietta's cells, a new entity rises. This new thing complicates matters. What does this new thing mean for science? Sometimes I think we forget that new scientific knowledge can be created. I hope no one is taking the term Frankenstein's creature to mean that Henrietta is a monster. I am using monster to mean ... something new ... something that is difficult to explain. Mary Shelley created *Frankenstein* in order to "tell a monstrous tale of science and education" (Weaver, 2010, p. 35). Shelley's vision of education "fosters an instrumental approach to education in which students at all levels are reduced to test taker and job seeking, proto-consumer while the fostering of the imaginative mind is neglected" (p. 35). Does that not sound hauntingly familiar? We are cranking out test takers each day.

Cartwright: How often do we take standardized tests as adults? Not a lot of real-life application.

Weaver: Are all of you familiar with Shelley's story, *Frankenstein?* Did you know it was about education … and not just a monster? *He looks out toward the audience.* Well, the creature … the creation… "overreached the human boundaries that separated humanity from the dangerous forces of nature" (Weaver, 2010, p. 35). Henrietta is part of a story that does at times seem like science fiction. When Henrietta was given immortal life through her HeLa cells a boundary was crossed. Henrietta is immortal through her cells. Her cells make her posthuman. "At the core the posthuman condition implies the merging of humans and machines in order to enhance or improve human capabilities" (p. 11). Cells removed from Henrietta's body were kept alive through techniques developed to grow cells outside the body. Those techniques along with George Gey's "roller tube" technology created the possibility for Henrietta to enter the post human realm (Landecker, 2007, p. 112).

Mary Shelley's work allows for an opportunity for us to contemplate understandings of science and education. I think the story of Henrietta's immortal cells can have a similar application. "Traditionally, scientists and science teachers see little value in understanding or studying the past. The only thing that matters is the present state of theory and practice within the natural sciences" (Weaver, 2010, p. 38). Why study history in science class?

> The history of science, for Shelley and most scientists, then is a waste of time, a luxury of time but not a necessity for conducting responsible science. Most importantly for Mary Shelley, the history of science is dangerous. Only contemporary science is enlightened and the type of science necessary to create scholar scientists while the past only hides dark secrets that are better left buried in time. (Weaver, 2010, p. 38)

What if Henrietta's story had remained buried? Instead, her story was resurrected. Mary Shelley ends her tale of Frankenstein with a cliff hanger. We don't know if the monster lives or dies (Shelley, 1818/1996). Today we are all wondering what happened to Henrietta. Can we tell this same tale, with Henrietta Lacks being the main character? Everything comes down to what stories can do for science and teaching … And here we have our own story that has been unfolding all day.

Appelbaum: Speaking of our story, I think the CNN has an update. I am going to put it up on the screen.

Polly Merase: This is Polly Merase for the Cellular News Network, reporting live at The World Alliance of Science and Science Education Symposium. After a day of excitement at Johns Hopkins University, there seems to be no sign of a person thought to be impersonating Henrietta Lacks. The BioArt exhibit was destroyed and a wax mannequin of Henrietta was stripped of clothing. Multiple reports of a woman entering lecture halls and entering discussions with scientists and curriculum theorists are surfacing as we continue to investigate this story. Many report they are not sure what they saw today. As we weave this story together, we know one thing ... there is no trace of Henrietta Lacks here in this exhibit hall ... Scientists on staff here say it is amazing that every cell from the art exhibit is missing. Considering HeLa's contamination history, there should be HeLa cells growing all over the exhibit hall. We will keep you posted on this story.

Enter Hannah Landecker, carrying a piece of paper in one hand and her tablet in the other.

Wald: Hi Hannah, I had almost given up on you.

Landecker: I am so sorry I did not get here sooner. Today has been very chaotic. Supporters of chaos theory in education should be reveling in seeing how our conference on science education is playing out. I have with me the list created in the philosophy session this morning. The note taker titled this...." What Science Education Needs." I thought you university professors would like to see this. Education often looks to researchers at the university level to introduce pedagogy that ascribes to current and past philosophies in education. I will just read it out loud. I am very curious to see if you all have discussed any of these issues. Who is keeping the list for this session?

Blades: No worries, Hannah, I have been quietly keeping notes. Let's see how they compare.

Landecker: Okay, here goes:

1. Basic laboratory skills of inquiry/not strict method

Cartwright: Put a check by that one. We discussed the scientific method being too confining. This is a problem in physics.

Blades: In biological science, teachers often have students repeat experiments with confined endings. Students need more opportunities to inquire and know they can create new knowledge.

Landecker: Sounds good. What about

2. Science experimenting with possible open-ended outcomes. Then add connections to real life, including economics?

Weaver: Commodification of humans, human cells, and human organs ... check. Maybe put two checks. So much of science policy and scientific research rests in the power of those who hold the money.

Landecker:

3. Science connections to culture

4. Science connections to humanity

Lather: We must remember that all the stakeholders in science are human, and we must make sure we take care when telling their stories.

Landecker:

5. Science connections to other subjects (Interdisciplinarity ... history, humanities, mathematics, anthropology, sociology, political science, and more)

Appelbaum: I would think Bruno Latour would be very happy to see such an interdisciplinary science. "Science pedagogy should pay attention to the importance of relationships among science, technology, society, and human values, and the importance of discourse and communities of inquiry in the classroom" (Appelbaum, 2001, p. 113).

Landecker:

6. Science connections to constant change

7. Science from a feminist perspective

Weaver: By working to include a feminist perspective in science, students of science see science "[constructing] a world where the knowledge [of women] and [their] experiences are valued" (Weaver, Anijar, & Daspit, 2004, p. 34). "Women give legitimacy to alternative ways of knowing" (p. 34).

Lather:

8. Science is not just for scientists, it must include all participants

9. Science must incorporate discourse in order to solve problems

The entire panel continues to nod as she proceeds through the list.

Blades: So far, so good.

Landecker: And finally, yet importantly, the group made a unanimous decision that Henrietta Lacks should be number ten.

Weaver: They put Henrietta on the list? Don't get me wrong, I have no problem with it … but I thought my own bias about her being included in science might be limited to a smaller audience. And the philosophers added Henrietta. Not much of a surprise if you really think about it. Science and science education need to value all its participants. (*He notices that Mandy looks as if she wants to speak.*) Mandy, did you want to add something?

Hoffen: I have often wanted to ask if you thought Henrietta could be accepted in a posthuman world. We all know she was an outcast in terms of humanism. With all our discussion today, it sounds like Henrietta has been invited into the house. John, remember how you used "Toni Morrison's imagery of Jacob's empty brick house in *A Mercy* to suggest we do not know who will count as posthuman" (Weaver, 2015, p. 182)? Morrison told the story to describe "that no matter how different Jacob and his progeny were from the D'Ortegas of the world, humans were still unfit to dwell in the home fit for the cultured and civilized" (Weaver, 2010, p. 182). Philosophers and curriculum theorists have included Henrietta in the conversation today, but what about the rest of the world? What about humanity as a whole? You tell us that humanity is "not fit to inhabit a posthuman world because we are still too barbaric toward one another selling everything for the right price" (Weaver, 2015, p. 182). Academic meetings of like minds can be like mountaintop religious experiences. We are all celebrating Henrietta Lacks here, lingering in this place and time—all likeminded. Nevertheless, just as Jacob was unfit to abide in the house, I worry that science will continue to see Henrietta as unfit to dwell in science classrooms, as herself.

Weaver: I must go back to Toni Morrison myself to respond. I can see a resemblance to Henrietta in the closing of Morrison's book:

> I am holding a light in one hand and carving letters with the other. My arms ache but I have need to tell you this. I cannot tell anyone but you. I am near the door and at the closing now. What will I do with my nights when the telling stops? (Morrison, 2008, p. 189)

Henrietta will only be known if her story is told. For her to dwell in those science classrooms, her story must be told.

Blades: Telling the story of Henrietta in those science classrooms will make Henrietta immortal, just as her HeLa cells are immortal. Could this possibly lead to an immortal love for learning?

Appelbaum: I want to assure everyone that I am all for Henrietta becoming as immortal as her cells. I think for those who are outside of this circle, we must make sure we frame our work in a manner that it will be suitable to a very large audience. We must make sure everyone understands how this story will be applied as alternative pedagogy in classrooms. I want to emphasize a few guidelines to make sure there is no confusion:

> In the alternative pedagogy, we search for an intimate knowledge of what interests our students bring to the classroom; these interests are the grounding of science as practice. New questions emerge. As they surface and flutter, we express anguish: that these new questions keep in mind the old ones and avoid a repetitive competition for rightness in favor of a dialogue about efficacy. In this dialogue we hope for attention to the conflicts that we have identified in contemporary discourse and practice. (Appelbaum & Clark, 2010, p. 597)

Weaver: We all know that we cannot reduce everything science needs down to a Letterman Top 10 list. If we are able to change the thinking and "break the grip of those stifling binaries that infest the minds of traditional enlightened science, we can create dynamic classrooms that turn science into a process of discovery and intellectual debate rather than a process of replication and rote memorization" (Weaver, 2001, p. 21).

Appelbaum: One other concern. Should Henrietta Lacks be the only name on this list? Don't get me wrong. In this time and place it seems to make sense to add Henrietta. Let me assure you I am well in favor of adding Henrietta. However, there are many others that need to be included as well. I suggest a wording change. Maybe say—Henrietta Lacks, her story, and the stories of unknown others?

Lather: That opens the boundaries. This wording ensures other opportunities for discourse to be introduced in the classroom.

Appelbaum: David, you used the word critique a few moments ago. I think we must continually critique "the narratives that are presented in school science" (Appelbaum, 2001, p. 125). We should also teach this critique process to our students. There are several questions that we must ask continually when evaluating what stories can become part of science:

1. Who benefits from this version of the story?

2. Whose prior knowledge and cultural experiences are best matched to the most important principles of the lesson? And whose are excluded?

3. How will I get students to ask and answer these questions themselves?

4. What (community) action projects will ask students to participate in local political processes? (Appelbaum, 2001, p. 125)

With these goals in place, we take a story to a new level. A third space for discourse is created.

Blades: Two groups, two different rooms ... and I think we have mentioned all of these in our discussion.

Weaver: We have not said much about power. Maybe some of these ideas need to be made more specific in their relationship to power. "At its heart, science education is not only about the teaching of science but the complexity of knowledge" (Weaver, 2001, p. 17). (*He looks toward his graduate student Mandy and speaks to her directly.*) Mandy, you look like you want to add something.

Hoffen: As soon as you said complexity of knowledge ... I wanted to say— Complexity despite policymakers reducing science to simplistic shopping lists of standards! Complexity despite educators being viewed as technicians and robots. A list is okay if it is just used as a guideline. However, the current measuring stick for success in science is based on the student's ability to regurgitate the information. "Bulimia," as Dr. Blades said earlier (Blades, p. 71). Predetermined standards and predetermined outcomes.

Weaver: I agree that science has too many boundaries. Simplified science reduces opportunity for change. "Complexity of knowledge and life and the importance of making sound public policy decisions that are not done for ideological or profit motives but for sustaining of a quality of life" (Weaver, 2001, p. 17). Now to create a "viable alternative to traditional enlightened science, [we] will have to invent a pedagogy that offers alternative ways of seeing nature, science, and the world" (p. 17).

Blades: We have contributed to a rather sizable discussion concerning pedagogy. It is interesting to me that our contributions all involve stories of some genre. Through science fiction, allegory, fables, nonfiction ... science can be expanded. Introducing stories to science provides connections that allow students to see science together with other disciplines such as literature, economics. When students see science connected to real life situations, they can form their own critiques. They may see or suggest that science and science education need to change.

Lather: I think it is rather poetic that we are having this discourse amid a story unfolding—a story within a story. Do we know what happened to Henrietta? Was Henrietta here at Johns Hopkins—walking around the meeting? Who was that person in the lobby asking for directions to the public ward?

Cartwright: Henrietta walking around the meeting is not the only story on the table today. Let us not forget the stories that we bring to the table today. Stories that tell of our unusual encounters on the way to this meeting. They seem of the supernatural sort ... not anything that can be explained by physics.

Weaver, Blades, and Appelbaum hum the theme to The Twilight Zone.

Cartwright: (*smiling*) "What we need to understand in order to understand the way scientific laws fit the world, it is the relationship of the abstract to the concrete, and to understand that, it will help to think about fables and their morals" (Cartwright, 1999, p. 36). If we really want people to understand the complexities of science, we must attach the science to something that an individual can relate too. "Fables transform the abstract into the concrete and in so doing, I claim they function like models in physics" (p. 36). I have a question. I want to ask if the "relationship between the moral and the fable is like that between a scientific law and a model. If they only apply in very special circumstances, then perhaps they are true just where we see them operating successfully, in the artificial environment of our laboratories, our high-tech firms, or our hospitals" (pp. 36-37). Personally, "I welcome this possible reduction in their domain, but the fundamentalist will not" (p. 37). If not fables and morals, how about considering the use of allegory? "Allegories say not what their words seem to say, but rather something similar" (p. 39).

Blades: "Allegories are a method of presentation in which an idea, person, or event stands for itself and /or something else" (Blades, 1997, p. 128). "Allegories have provided throughout time a way for authors to introduce critique to discourse while avoiding premature closure of meaning" (p. 128). Some of your journeys to this meeting were spectacular. "Think about the exploits of Don Quixote, discoveries of Gulliver, Christian's journey as a pilgrim or obsession of Ahab to find the Great Whale are but a few of the masterful allegories used by authors to present the public an invitation to discuss something else" (p. 21). Although *The Immortal Life of Henrietta Lacks* is not an allegory, the same purpose is fulfilled. This book provides a wonderful example of an author presenting the public with an invitation to discussion; Skloot invites all the readers to participate in discussions about racism, sexism, segregation, economics, and ethics, just to name a few.

Lather: Struggle is a common theme. Allegory allows us to tell a story privately. An allegory "can speak quietly, with respect for all that it means to tell the stories of people ... without making a spectacle" of their private struggles (Lather & Smithies, 1997, p. xiii).

Weaver: I agree that allegories have a definite place for creating discourse in any setting. One could say the same thing about science fiction. If one were reading Rebecca Skloot's book, out of context, one might think it is science fiction ... immortal cells, cell lines, cells going to the moon. As curriculum theorists and educators, science fiction allows us to "push boundaries" (Weaver & Anijar, 2001, p. 15). Science fiction can become "our central metaphor, negotiating previously uncharted territory to reconceptualize epistemological, axiological, and ontological constructions" (p. 15). Science fiction allows "teachers and students to speak on their own terms" and break away from "the strict codes" created by academia and government (p. 15). Today's events all have elements of science fiction. Do any of you think Henrietta could have risen from the vat of HeLa cells? I know Mandy is thinking that …

Appelbaum: That is an idea we should entertain, if we are going to be the open-minded curriculum theorists that we are.

Weaver: Let's say it did happen … our Frankenstein creature Henrietta is here walking the halls of Johns Hopkins again. Where would she go?

Hoffen: She would have to find somewhere safe, some place she was accepted, a place where people believed that she was real, and felt that she deserved to be treated with respect. She could come home with me. My school has been nicknamed the Henrietta Lacks School in our district of six high schools. The students in my school would accept her. The students talk about her each year when we celebrate her life.

Wald: Like I said before Mandy, you have imagination! Please let me know if you find Henrietta. As we close today, I must say that this has been a most excellent session. I am glad there are others that must carry the responsibility of finding the woman who has haunted our proceedings today.

Landecker: It seems ironic to me that "cell lines are made to stand in for persons in the first place; they function in laboratories as proxy theaters of experimentation for intact living bodies" (Landecker, 2007, p. 54). Now we may have seen that in reverse … a real live woman … resurrected from the cell line, representing her cell line in our sessions.

Wald: Mandy, I think your imagination is wearing off on Hannah!

POSTSCRIPT: ALTERNATIVE PEDAGOGIES

As a science educator in today's public schools, teaching a standardized curriculum, I want to take the basic required curriculum and expand it so that it "allows for a multiplicity of possibilities and alternative pedagogies" (Weaver, 2001, p. 20). Weaver suggests that all teachers must switch the dial from the required process of rote memorization to a "process of discovery and intellectual debate" (p. 20). Too many times teachers are simply the givers of answers in order to expedite teaching and gain results expected by the state or district office. In an age of accountability for teachers, brought about originally by No Child Left Behind, and now Race to the Top and Teacher Keys Evaluation System, teachers have been forced to focus their instruction on standardized testing. Standardized tests are used to evaluate teacher performance, as well as student performance. Test review becomes a priority in classrooms. With added time devoted to test review, and mandated benchmarking, material is condensed in a manner that assures objectives are taught in a condensed manner. Teachers in my building report a decreased amount of time for instruction that emphasizes critical thinking and even time for laboratory exercises.[1] Gayler (2005) looks at standardized testing specifically in Maryland and Virginia. In these two states "teachers have revised their instruction to emphasize topics and skills likely to be tested and to spend more time on reviewing information and test-taking skills" (p. 4). He goes on to say that students and teachers report "that instruction has become too focused on reviewing discrete facts, with little time for discussion, and in-depth-learning or creative lessons" (p. 4). In such a setting it is the "students [that] are being left behind as teachers push ahead to cover all the topics on district curriculum maps and pacing guides" (Gayler, 2005), p. 4).

Good teachers will prepare the students for the mandatory testing. Regardless of whether we agree with this policy, we want our students to be successful on these standardized assessments. The assessments influence their grade, and at the high school level, grades impact students' futures.

It is a bit of a juggling act in my classroom to make sure standards are being taught at a level that requires critical thinking, and at a level where inquiry and learning to ask questions is the focus. Having the ability to take a standardized test does have some merit. High school students take SAT's and ACT's in order to get into college. High school students take the ASVAB in order to gain entrance into the military. Eventually we hope many students will sit for licensing exams in fields of medicine, nursing, teaching, plumbing, electrical work, cosmetology, and many others. In our day-to-day lives these testing opportunities do not have much application. I have not taken a standardized test since I took the Graduate Record Exam to gain entrance into my doctoral program. In my classroom this past spring, during a 6-week period, some students endured over 17 different standardized assessments. The students' exhaustion was obvious in their demeanors, and their physical well-being.

Time in science class should not focus on testing alone. Science curriculum should be rich with philosophy of science, history of science, activities that involve inquiry, and stories. Students can use stories to relate their results and a new way of thinking with familiar information and prior knowledge, essentially bridging the gap between the known and the unknown. Boundaries of science should be expanded to include all participants—those researching and creating new knowledge in science and those impacted by this new knowledge. Can we begin by removing science and scientists from the pedestal upon which they were placed during the Age of Enlightenment? Weaver (2001) tells us that society can no longer "shroud science behind a cloth of god-like importance and ability" (p. 6). Weaver (2001) also tells us that "the notion that the public is too ignorant to understand and take part in major scientific policy and decisions" must end (p. 17). Too often students of science think that science is out of their reach, yet we all use science in informal ways. For example, one might observe that he or she is not sleeping well on Thursday night. Upon further reflection and a quick effort to Google the caffeine content of Starbucks coffee, one might discover that the cup of decaffeinated Starbucks coffee consumed while picking up a few groceries after work might contain enough caffeine to disrupt sleep. Science is experienced daily, from our observations to the medications we take to control our allergies or high blood pressure. We know that beyond standardization, science education must be socially and culturally relevant.

Deboer (1991) tells us "Socially relevant science education was promoted in one form or another by most science education leaders during the first half of the twentieth century with calls for scientific literacy and with the creation of a science-technology-society approach" (p. 234). "It is curious then, that science educators have had such a difficult time convincing classroom teacher of the merits of a socially relevant approach" (p. 234). Teachers from the early twentieth century tended to spend too much time "on the study of science for 'its own sake' and not enough time on socially relevant themes" (p. 234), a practice continued by present-day teachers. Deboer (1991) presents three reasons to support teaching socially relevant science. "The first is that by teaching science in the context of what is already familiar from daily experience—newspaper, television, and magazine accounts of nuclear power plants, environmental pollution, recycling and the ozone layer, for example, or about objects in and around the house such as electronic devices, the automobile, and household appliances—the student is motivated to learn the science that relates to those daily experiences" (p. 235). The curriculum becomes connected to the student.

When science is relevant, "worthwhile outcomes result" (Barton et al., 2003, p. 118). These authors report that

> Students of science begin to recognize and exercise their voice and autonomy; they learn to become agents of change in their own lives and within

the disciplines of science, using their authority to challenge the traditional cultural practices of science and education. (Barton et al., 2003, p. 118)

When learning experiences make a connection with the student or the student culture, the information becomes part of the student. "Resonant learning experiences are meaningful and empowering" ... by creating, while "responding to their own questions and needs, science takes on a personal relevance that is not something that had to be learned for a test or project" (Barton et al., 2003, p. 118). Students intersect with science on multiple levels: "Power, access to knowledge, equity, social justice, and culture play a role in how youth experience science, but also in how youth respond to difficult situations in order to create a practice of science that has power and meaning in their lives" (p. 158). The intersections become part of their life, their story. "Stories help us understand how science, schooling, and society might intersect in science education settings to help build a more socially just, critically informed, and sustainable society" (p. 158).

Deboer traces science education from the early 18th century to the late twentieth century. Over time, social relevance has gained ground over traditional rote memorization. Deboer (1991) tells us "the true purpose of education was to prepare people to deal with ... socially relevant questions—to equip them for the age in which they live" (p. 3) He tells us that if we look and "see what has been lost and gained by each major shift in the past," we can "head with greater vigor in the direction of social responsibility and socially relevant instruction in science" (p. xii). In the next chapter we will see how students perceive the story of Henrietta Lacks. In Act VII we will hear student conversations about Henrietta's story, as well as their responses. Student conversations taking place in response to the story of Henrietta provide evidence of lingering. New ideas can arise from the space in between the story and the students.

NOTE

1.   See Heese (2015) for a thorough literature review concerning consequences of high-stakes testing.

CHAPTER 6

THEORIZING PRACTICE ~ PRACTICING THEORY

PRESCRIPT: DEBORAH LACKS AND HELA CELLS

On May 11, 2001, Henrietta's daughter Deborah, and her brother Zakariyya, met Rebecca Skloot at Johns Hopkins hospital. Today was the day they would see their mother's HeLa cells for the very first time. Their other two brothers could not make the meeting. One was working and the other one was going to see a lawyer about suing Johns Hopkins. Rebecca, Deborah, and Zakariyya would be visiting the laboratory of Christoph Lengauer. The scientist met them in the lobby with a smile and outstretched hand, wearing blue jeans and a plaid shirt. Looking at Deborah, He said, "It must be pretty hard for you to come into a lab at Hopkins after what you have been through. I'm really glad to see you here" (Skloot, 2010, p. 261). He explained that they would start in the freezer room, so he could show the group how Henrietta's cells were stored, and then he wanted them to see the cells under a microscope. Christoph opened the freezer. Deborah was in awe that she was finally face to face with her mother's cells. Zakariyya was very quiet. Christoph showed the group a vial full of red liquid complete with the "letters H-e-L-a written on its side" (p. 262). Christoph explained that each vile contained millions of cells. If the cells stayed in the freezer, they could be thawed out to start growing again. He told them that the cells could stay in the freezer forever. "Fifty years, a hundred years, even more—then you just thaw them out and they grow" (p. 262). Growing cells in a laboratory is strictly science. Right?

Mandy Hoffen and a Conspiracy to Resurrect Life and Social Justice in Science Curriculum With Henrietta Lacks: A Play, pp. 139–167
Copyright © 2021 by Information Age Publishing
139

Christoph Lengauer's research centered on HeLa cells. He became very passionate about Henrietta's story after reading an article that Skloot wrote for *Johns Hopkins Magazine*. Christoph worked with "florescence in situ hybridization." FISH allows a technician to "paint chromosomes with multicolored florescent dyes that shine bright under ultraviolet light. To the trained eye, FISH can uncover detailed information about a person's DNA. To the untrained eye, it simply creates a beautiful mosaic of colored chromosomes" (Skloot, 2010, p. 234). When Christoph learned about Henrietta's story, he framed for Deborah one of his HeLa cell paintings. "It looked like a photograph of a night sky filled with multicolored fireflies flowing red, blue, yellow, green, purple, and turquoise" (p. 234). Christoph sent the print to Rebecca Skloot. Rebecca presented the print to Deborah when they met for the visit to Hopkins. Christoph opened his laboratory door and allowed Henrietta's family to enter. A space was created between the science of HeLa cells and the real people involved in the story, Henrietta's family. A new story evolved. Christoph reached out and Deborah gained learning and understanding about her mother's famous cells that had eluded her throughout her entire life. This excerpt of the story from *The Immortal Life of Henrietta Lacks* is a wonderful example of curriculum in the making. As I read about Deborah, Henrietta's youngest daughter, I quickly noticed that she is a self-motivated learner. She spent all her life teaching herself about cells. She would ask questions. She would read anything she could find about her mother's cells. Real-life experience provides the spark that can make a school curriculum come to life. By adding personal experiences with science as common ground for discussion, all stakeholders can become part of the conspiracy that can enhance a standardized curriculum.

ACT VI

The action leaves Johns Hopkins and returns to the school where Mandy Hoffen works. She is attending a meeting with a curriculum director, a university professor, and another teacher. The meeting is plagued by one interruption after another. Participants in the conversation will leave their meeting room and observe student conversations in a construction classroom, and a biology classroom. Student responses will be featured in this chapter.

Act VI: Henrietta Goes to High School

Characters in order of appearance: Doris Bradley, curriculum director; Krystal Sonnenshein, school receptionist; Alexis, high school student;

Lydia, high school student; Mandy Hoffen, high school teacher; Katherine Driscoll, university professor; Delores Pequod, high school teacher; Wayne Cooper, construction teacher; Matthew, construction student; Anthony, construction student; Jhaymia, student presenter; Kyle, student presenter; Charles, student presenter; Mryrtis. Robinson, teacher.

Setting: Act IV is set in a large suburban high school in the southeastern United States. An educational researcher/curriculum director and a university professor meet with two science educators to discuss the current state of science curriculum at the school. Students provide several interruptions during the meeting. The action begins in a media center conference room, and then moves to three separate classroom settings.

Act V1 Scene I: A Noisy Meeting

Doris Bradley navigates her rental car through the school parking lot. She enters the double doors. Posters that say Wear Red October 4 to Honor Henrietta Lacks cover the entryway. The receptionist behind the window is wearing a bright red shirt and shiny red Mardi Gras beads. She has a sticker and red ribbon sticking to her name tag.

Bradley: Hi, is it Ms. Sonnenschein? What does your sticker say?

Sonnenschein: Yes, it is! The sticker is the book jacket for *The Immortal Life of Henrietta Lacks* (Skloot, 2010). You know, the 2010 best-seller written by Rebecca Skloot. Our biology students read this book when they study cells. We are wearing red today to cell-e-brate the life of Henrietta Lacks. Get it, Cell-e-brate. You know like cells are the basic unit of life.

Bradley: Ah, yes. Clever play on words. How exciting!

The receptionist smiles, hands Bradley a yellow visitor sticker, and gives her directions to the Media Center conference room. As she enters the school commons area, she hears music—a Duke Ellington tune. She sees a booth set up close to the Media Center door. Everyone in the booth is wearing red T-shirts that say—Not just HeLa—Henrietta Lacks. They are stacking up bookmarks, untangling red beads, and positioning red balloons. A young woman approaches Bradley.

Alexis: Would you like some beads? I can give you some if you are willing to tell Henrietta's story.

Bradley: Sure, I will take some beads, but who is Henrietta?

Alexis: Henrietta lived in 1951, when women were nobody, and Black women were even less than nobody. She got sick and had to go to Johns Hopkins Hospital; any other hospital would have let her die in the parking lot because she was a Black woman. The doctors diagnosed Henrietta with cancer. Mary, a technician from George Gey's Lab, went to the operating room and took a sample of the tumor cells. It was her job to try to grow the cells outside the human body. No one had ever done this. Usually the cells would just die, but after two days, Henrietta's cells were growing like crazy! The cells were used to develop the Polio vaccine. Because we could grow these cells outside the body, we learned how cells work. Many medications and cures for diseases were discovered because of Henrietta's cells (Skloot, 2010).

Bradley: How is it that I have never heard of Henrietta?

Alexis: They kept her name a secret. The doctors did not want her name to be released. I think they were afraid the public would find out she was Black (Skloot, 2010).

Bradley: This sounds like a great story.

Lydia: Our teacher has us research the story and find a part of the story that we feel is important. Then we create a product to tell the part of Henrietta's story that we pick. We must think about how that part of the story is important to science. Here comes our teacher now. Good morning Hoffen.

Hoffen: Good morning to you too, Lydia. Hi, Alexis. Welcome Dr. Bradley, it is nice to see you.

Alexis: Dr. Bradley, are you here for our Henrietta Lacks Cell-e-bra-tion? Get it … Cell … Cell … like the cells in your body.

Brandley: I am hoping to stay Alexis. But right now, we have a meeting.

Hoffen: Alexis, can you walk Dr. Bradley to the conference room? I will be right there.

Alexis shows Dr. Bradley the conference room. Katharine Driscoll, a university professor and Mrs. Pequod, a teacher at the school, greet them.

Alexis: Hi Mrs. Pequod. Are you coming to the program today?

Pequod: No Alexis, I will not be at your program. You need to get to class.

Alexis: I do need to get back to the booth. I hope to see you later Dr. Bradley.

Driscoll: Hello Dr. Bradley, thank you both so much for volunteering to be part of this conversation. Your entire science department seems very enthusiastic about developing a program to bridge the gap between educational theory and teaching practice. As educational researchers, we need to see if our theoretical perspectives can be applied to actual classrooms.

Pequod: Wait a minute. You think I want to be part of this theory-meets-practice research project? (*She speaks very gruffly to the group.*) I want to make sure you understand that we do not want to be part of your program. Our school is fine just like it is.

Mrs. Hoffen rushes into the room dressed in red.

Hoffen: Hello everyone, I am sorry I am running late.

Driscoll: We are so glad you are here. We were excited to learn that our meeting is on the same day as your event. We want to see this Wear Red Day to Honor Henrietta Lacks firsthand. Our panel is hoping to see real-life examples of educational theory meeting educational practice.

Hoffen: Wonderful. Feel free to have some lunch, enjoy the music, and then attend our program in the auditorium. Mrs. Pequod, I do hope you bring your classes today.

Pequod: Are you kidding? We must finish our benchmark and review for the biology end of course test.

Driscoll: Let's start. I am Katherine Driscoll. My department at the university is interested in how theoretical information taught to teachers can be applied to actual classroom settings.

Bradley: I am Doris Bradley, a science educator and educational researcher. Currently, I am serving as this district's curriculum director. I want to see science taught in a manner that involves social justice, democracy, and human rights. When we work with students to "create relevant science, many worthwhile outcomes result" (Barton et al., 2003, p. 118).

Hoffen: It is so nice to meet you Dr. Driscoll. My name is Mandy Hoffen. I am a science teacher here.

Pequod: Excuse me, did I hear you right, Dr. Bradley? Science has nothing to do with justice, democracy, or human rights. Science is the pursuit of truth and understanding through a trial of experimentation and verifiable and repeatable results.

Driscoll: Mrs. Pequod, part of our work at the University involves an extended definition of science. I will explain my views of science and science education as the conversation unfolds today. First, I have a question: Why do we have jazz today?

Hoffen: Henrietta and her sisters would get together and dance to jazz music on Saturday nights. They loved the tunes of Benny Goodman, Louis Armstrong, and Billie Holiday. They would move all the furniture out of the living room and onto the lawn and dance until dawn inside the house (Skloot, 2010).

Pequod: You just go on and on don't you. I probably have 50 emails about her HeLa day on my computer right now.

Hoffen: She was not just HeLa. I prefer her to be called Henrietta Lacks.

Driscoll: Let us continue with our introductions. Mrs. Pequod, what can you tell us about yourself?

Pequod: It's Delores, Delores Pequod. I have 29 years in, and I will be surprised if I make it to the 30-year mark. I teach whatever curriculum they tell me. There is no time for stories or jazz (*she glares at Mrs. Hoffen*). We have standards to teach, a curriculum map to follow, and end of course tests. I had a 91% pass rate last year.

Driscoll: Thank you all for being here. Mrs. Hoffen, why do you use stories in your science classroom?

Hoffen: I feel that it is a way to show real world connections. We must allow all aspects of our world "into our method" (Grumet, 1990, p. 107).

Bradley: I taught biology just like you two. However, I see a tremendous need for real world connections.

Pequod: My textbook has all the stories I need, and it is aligned to our standards.

Bradley: Schwab (1978) tells us that

> To employ only one doctrine as a principle will give a biased view of the nature of science, and to teach a single doctrine

should be to the student only misleading, confusing, or both, because no single doctrine is more than a partial statement. (p. 72)

Teachers not only have to consider science; they must consider theoretical views of science education. Curriculum theories and teaching and learning theories "cannot alone tell us what and how to teach, because questions of what and how to teach arise in concrete situations loaded with concrete particulars of time, place, person, and circumstance" (Schwab, 1978, p. 322).

Driscoll: Theory and practice, the two must come together. A new entity must be created between the two.

Hoffen: Science curriculum needs rethinking, revisioning, revising, or maybe reconceptualizing. We must linger together and give this serious thought, considering all participants. Dr. Driscoll let me give you a quick update about what is going on here, in this school district. We have performance standards that dictate what teachers should teach. I have provided each of you a copy. The desired learning is to result in predetermined prescribed outcomes. A strict curriculum map dictates the entire teaching schedule. This map not only tells when topics should be covered, it determines the sequence, and time allowed per unit. This map is packed full of science objectives. If students have difficulty with an objective, it is virtually impossible to find class time to address remediation. When students are not being tested, they are being drilled on how to be successful on standardized tests. Our district expends large amounts of energy on staff development to teach teachers the right formula for teaching, so the curriculum map can be efficiently followed, and students can be successful on standardized tests.

Driscoll: I feel that schoolteachers have a very difficult job. How can they work "amidst the distraction of government [intervening] in the intellectual lives of teachers and students" (Pinar, 2004, p. 208)? First, it was No Child Left Behind. Now schools must meet the guidelines sketched out in Every Student Succeeds Act (ESSA). ESSA holds states "accountable for focusing resources on low-performing schools and traditionally underserved students who consistently demonstrate low academic performance" (Alliance for Excellent Education, 2019). Both involve standardized testing. Unfortunately, standardized tests are just one moment in time. They are no more than a snapshot of what a child can do at one exact moment. An objective list can never be all-inclusive. I think it is correct to say, "Education, experience and life are inextricably inter-twined" (Connelly & Clandinin, 2000, p. xxiii).

Hoffen: My students pay attention more when they are interested in what is going on in class. I find that allowing them to be stake-holders in the daily plans, and subjects to be discussed, makes them take better charge of their own learning. Cells, for instance, is such an abstract subject for students. I invite students to view cells using stories. We begin by studying Anton Von Leuwenhoek and his invention of the microscope. Then, we add Robert Hooke to the story of cells. Who saw cells first? This is a tough question to answer. Their textbook tells a story about Hooke being the first to see and name cells. Students are asked to begin researching this topic on their personal devices. Then the arguing ensues. Hooke was first. No, Leuwenhoek was first! The stories help students make connections. These connections may deal with students' interests, or the value of the students' cultures that they bring to the classroom. There are relationships to consider. Curriculum is not exactly "neat and tidy," it is about the "interrelations between students, teachers, and texts" (Morris, 2001, p. 2011). Curriculum is a collection of noise in the classroom.

Pequod: Yep, with 32 students talking at once, and that intercom in the ceiling.

Hoffen: I agree Mrs. Pequod; it does get noisy. But, when I said noise, I was talking about the interruptions that disrupt our thinking. Think about a group having a discussion. Someone says one thing and the discussion is carried in multiple direc-tions. Then someone else's thinking is sparked by a previous comment. A chain reaction of sorts is activated in the class-room. Before you know it, everyone is talking, discussing, and the conversation is growing. New ideas are created.

The conversation is interrupted by a knock on the conference room door. Mrs. Hof-fen walks to the door to find Alexis and Lydia, all smiles.

Alexis: Mrs. Hoffen, we had to bring this to you. (*Alexis is holding a black t-shirt covered in cotton balls and glow-in-the-dark paint. Imme-diately, Mrs. Hoffen knows that the shirt is covered with HeLa cells.*)

Hoffen: Wow people ... this is wonderful ... (*She immediately pulls the shirt over her head.*)

Pequod: (*Speaking under her breath.*) But isn't it Wear Red Day?

Lydia: The paint on our shirts is still drying. Mine has. (*Mrs. Hoffen interrupts.*)

Hoffen: Girls this is incredible … you have made my day. I am somewhat busy right now with our guests. Thank you so much.

The girls run off. Mrs. Hoffen closes the door.

Hoffen: I am so sorry about the interruption. The girls are so excited about our day.

Pequod: Noise! Interruptions! How is anyone supposed to be able to get any work done around here? Why are those two girls allowed to run all over the school? Why are they not in class?

Bradley: That is not just a HeLa shirt. This young lady shared the story of Henrietta Lacks this morning when she gave me these red Mardi Gras beads. She is beginning to "recognize and exercise [her] voice and autonomy" through your project (Barton et al., 2003, p. 118). Wear Red to Honor Henrietta Lacks helps your students "learn to become agents of change in their own lives, and within the discipline of science, using their authority to challenge the traditional cultural practices of science and education" (p. 118). Both girls are proud to be learning science and teaching others about Henrietta Lacks. The girls are engaged "in relative science" (p. 119).

Driscoll: Thank you, Doris, for reminding us not to view the girls' visit as just an annoying interruption. Mrs. Hoffen, can you share with us the significance of your shirt?

Hoffen: (*All smiles, she picks up a dry erase marker and proceeds to the white board.*) You see, this is a cell membrane. She draws a shape like the ones on her shirt. Most of what we can see with a basic microscope is a cell membrane and a nucleus. The cotton balls represent the nucleus.

Driscoll: Did the girls get cotton ball happy? There cannot be 2–3 nuclei in each cell.

Bradley: I can answer that one. Cancer cells have more than one nucleus in each cell.

Hoffen: Yes, very good. I am so proud of them. We read about cancer cells in *The Immortal Life of Henrietta Lacks*. Each student must create a product to tell part of the story … the girls decided to make shirts. They were drawing rough drafts of their shirts earlier in the week, but the giant HeLa cell shirt was a surprise! I am so excited!

Pequod: That is all fine and good, but can we get on with our meeting, minus the noise and the interruptions?

Driscoll: Yes Mrs. Pequod, we can continue.

Bradley: I have a quick question. Mrs. Pequod, do you feel the standards are insufficient?

Pequod: The standards are packed. Students have a lot of information to memorize. My fear is that the standards do reduce and water down science to a list of facts to be memorized by the students. And my other fear is that there is not enough time to cover all the curriculum. We must stick to the curriculum map.

Driscoll: Brian Heese (2015) reported that high stakes testing increases teacher stress and anxiety, decreases teacher morale, impacts teacher relationships with other teachers and with students, increases the amount of work required in the school day, and diverts funding away from instructional materials and to materials needed for testing and testing preparation. High stakes testing also influences the number of classroom assessments and the formats of how exams are written as well as how exams are weighted in a student's overall grade.

Bradley: All the science teachers in my district indicate that testing creates stress for all that are involved. My worry is that we are not teaching students what they need.

Pequod: That is why you must teach the standards, according to the curriculum map. Otherwise, the students will not pass the test. There is not time for stories.

Bradley: By oversimplifying science to a list of standards or facts and processes to simply memorize, "we block progress toward one of the most widely held, if impossible goals" of what I call "liberal science training" which is "to provide the student with an adequate picture of the world" (Schwab, 1978, p. 99). You must understand that "to give a simple picture of a complicated world is not to give the scientist's picture of that world. By so doing we make the picture not only inadequate (which it will always be) but false" (p. 99). I have a moral obligation to my students. We must take their education seriously.

Driscoll: With all that is happening in education, it would be easy for "teachers to resign their professional authority and ethical responsibility for the curriculum they teach" (Pinar, 2004). In education, the process begins with a commitment to students and the school or university. Pinar (2004) tells us that

> Whatever our fate—given our betrayal by government and by powerful professional organizations, the future is not bright—we must carry on, our dignity intact. We must renew our commitment to the intellectual character of our labor. We can do so first by engaging in frank and sustained self-criticism. (p. 9)

No matter how difficult the journey, teachers must be willing to make the commitment to become more intellectual. If we want to model learning in our classrooms, our students must see us taking an active role to reflect on our own learning. I am sorry. I get a little carried away when I start talking about curriculum theory.

Pequod: So, you are telling me that if I analyze myself, my learning, that I can begin to cure all the ills of education?

Bradley: No, I am saying there cannot be a simple monochromatic view. We must use reflexive thinking, and view things from more than one perspective, and purposely "looking through a succession of lenses," standardized education offers only one point of view, a one-size-fits-all education (Schwab, 1978, p. 325). Current education is characterized by a "linear sequential, easily quantifiable ordering system," combined with expectations of "clear beginnings and definite endings" (Doll, 1993, p. 3). We as educators must realize the potential of finding a "complex, pluralistic, unpredictable system or network" upon which to build a foundation for educational experiences for students and teachers alike if we are willing to expand linear thinking (Doll, 1993, p. 3). I want to bring attention to all the key players in this situation: the teachers, students, and curriculum. I propose a "new concept of curriculum" emerging as "new relations between teachers and students" (p. 3). This new relationship of curriculum, students, and teachers will be quite complex … "Less ordered and more fuzzy" (p. 3). Such a relationship cannot be complete without considering the life experiences of each person involved, as well as the real-life consequences and applications of the curriculum taught. As educators, we must find our place in this highly textured, woven fabric. Just as threads intersect in a woven fabric—the weaving of a curriculum is like a....

Hoffen finishes Dr. Bradley's sentence.

Hoffen: I know that quote, "fabric with brilliant threads throughout" (Krall, 1999, p. 5).

Bradley: Unfortunately, the beautiful threads, or "gaps, breaks, punctures are not only absent from the curriculum, they are seen only in negative terms" (Doll, 1993, p. 37).

Hoffen: I think teachers are afraid to break away from the standardized practices. All the stakeholders in education have potential to provide input into a new curriculum. Students' voices are being silenced, and learning prescribed:

> If students are not able to transform their lived experience into knowledge and to use the already acquired knowledge as a process to unveil new knowledge, they will never be able to participate rigorously in a dialogue as a process of learning and knowing. (Freire, 1970, p. 19)

Dialog is a necessary entity when searching for what is worthwhile.

Bradley: I would add that teachers' voices are also being silenced. Dialogue, discussion, and discourse are all necessary in finding meaning and personal connections in curriculum.

Driscoll: The teacher is the facilitator for what is worthwhile. The teacher ultimately must embrace intelligence, allowing students to leverage what they know they can successfully accomplish. As students develop this essential opportunity, their imagination, interest, and creativity allow them to create a love for their learning that may endure the travesties and injustices they face both in and out of the classroom.

Pequod: Excuse me. I personally see no problem with standardization. In addition, as far as the event going on in the commons today, I see no relation to science. Discussions concerning justice and democracy should be saved for social studies class. We have too many objectives to teach. Teaching is simple; we teach the standards, the students take a standardized test, we know that learning has taken place.

Bradley: I must disagree, Mrs. Pequod. We as teachers have the "potential for making science transformative in the lives of our

students, thus creating opportunities for students to have resonant learning experiences that are educationally meaningful
and empowering" (Barton et al., 2003, pp. 118–119).

*There is a knock on the door. Mrs. Pequod, closest to the door, gets out of her chair
and answers. A group of students shows up at the door of the conference room
dressed in red, wearing red Mardi Gras beads, and wanting to talk all at the same
time.*

Pequod: You kids must go to class.

Alexis looks past Mrs. Pequod and around to Mrs. Hoffen.

Alexis: Mrs. Hoffen, which was a bigger problem in the 1950s—sexism or racism?

Hoffen: Both were problems.

Alexis: I think racism was a bigger problem because it affected men
and women …

Lydia: I think sexism was a bigger problem, because women were
impacted by racism and sexism.

*Both girls are sporting red shirts with the name Henrietta Lacks across the front in
large puff paint letters. On the back of Alexis's shirt is a Venn diagram. The circle
on the left is labeled "Whites," the circle on the left is labeled "Blacks." Lydia's shirt
is very similar, but her Venn diagram is labeled "male" and "female" with a large
"1950" painted above.*

Hoffen: An author named bell hooks writes about the racism and
sexism she experienced while she was in college. She attended
college in what she called "the wake of a powerful anti-racist
civil rights struggle" (hooks, 2010, pp. 2–3). This was in the
50s around the same time Henrietta lived. Hooks (2010) tells
us "the outspoken sexism of [her] undergraduate male professors were even harsher than their covert racism" (p. 3). She
experienced racism and sexism at the hand of her teachers
"who appeared to derive their primary pleasure in the classroom by exercising their authoritarian power over [her] fellow
students, crushing [their] spirits, and dehumanizing [their]
minds and bodies" (p. 2). Alexis, can you "imagine being
taught by a teacher who does not believe you are fully human"
because you are Black (p. 2)? Ladies, this is a very complex
issue. I know you two are having a disagreement over trying to

decide who is right. Maybe both of you are. Unfortunately, sexism and racism still exist today. Both are major problems. bell hooks became a teacher. As a teacher, she believes that she must write about these problems from her perspective as a Black woman. There were not many Black women in her time writing about issues of civil rights. I think our focus must be on making others aware that racism and sexism still exist today. Which is exactly what you two are doing today by sharing Henrietta's story. If people do not know, nothing can be done.

Pequod: (*Clears her throat loudly.*)

Alexis: (*Whispers.*) Thank you, Mrs. Hoffen. We will Google her. I guess we will be on our way to class now.

Hoffen: I apologize once again for the interruption. Alexis and Lydia are proof that there is more to be considered than that student's grade on an end-of-course test. Science can be "difficult and forbidding, even intimidating and remains the province of experts. Few students ever achieve a personal understanding of science; few students ever really own the science they study in school" (Hodson, 1998, p. 6). That is unless the students find a way, a place, and a space that allows them to connect personally with the science.

Students and teachers need a space for teaching and learning. The learning that I desire to take place will not be something that can always be measured with a paper and pencil assessment. The learning will be evidenced by what my students carry with them outside of my classroom and into the world. The space we create will make time for students to develop their voice, their thinking, their relationship with the subject matter and the world in which they live.

Bradley: The girls' interruption was timed perfectly. This Wear Red Day as a part of science curriculum allows science to be "recast as something that youth create in the process of responding to their own questions and needs. It is not something that they have to learn for a test or for a project that they must complete to satisfy a teacher's requirements" (Barton et al., 2003, p. 118). The idea that you would allow the students to choose a portion of the story to tell, which gives them a space to ask questions of their own, is empowering to students. In this type of activity, "the process of doing science involves [students'] agency and authority in articulating their questions, and in constructing ways to respond to those questions" (p. 118). The girls both chose a question, and then chose how they would respond to that question.

Driscoll: The project creates a space for students to be more vocal. A space for teachers to be vulnerable. Every classroom could benefit from such spaces. Homi Bhabha was one of the first to write about in-between spaces. He wrote about these spaces in reference to cultural differences (Rutherford, 1990). Relationships naturally provide opportunities to create in-between spaces. As you introduce the story of Henrietta Lacks, you are opening opportunities for discussion of race. This is something that needs to be happening in the U. S. South daily. The story allows you to bring race to the forefront of the conversation. If you ignore your race, you become isolated. If you ignore the race of your students, they become isolated. In-between spaces created in the classroom, whether the space deals with cultural differences, relationships, or subject matter, will "flourish through the joined efforts of individuals who invent spaces for creating values for all so that all humans might live more robustly, develop capacities more fully, and become humane and educated in an increasingly diversified, complicated, and contested cosmos" (He, 2013, p. 63).

Hoffen: Cosmos. (*Speaking excitedly.*) Have you studied Alexander Von Humboldt? Humboldt and I share similar backgrounds in science, characterized by empirical data and strict method. Humboldt "came to believe that imagination was as necessary as rational thought in order to understand the world" (Wulf, 2015, p. 36). In a letter "Goethe encouraged [Alexander] to combine nature and art, facts and imagination" (Wulf, 2015, p. 38). This new thinking allowed Alexander to think beyond science and see the connection of science to the world and the peoples of the world. Humboldt, who can be described as a scientist, an explorer, and a bridge builder, spent his life traveling, observing, collecting, measuring, and connecting. Humboldt wished to create bridges between "peoples, disciplines, places, and historical eras" (Walls, 2011, p. 10). Humboldt hoped to "create a zone of exchange rather than domination" (p. 10). I want to create this kind of "zone of exchange" in my classroom (p. 10).

Pequod: Cosmos? I am here to discuss educational theory, and she is spouting off stories about an explorer whose accomplishments were forgotten long ago. So long, I have never heard of him.

We can barely get these students to memorize the parts of the cells, and you want to make a million connections.

Hoffen: Alexander von Humboldt predicted problems with "deforestation" and "climate change" years ago (Wulf, 2015, p. 58). Tell me that is not relevant to science education and society.

Driscoll: We keep dancing around the idea of in-between and so-called bridges. Let us get specific. How will you build that bridge between your students and science?

Hoffen: The first set of bridges we need to build will involve relationships between teachers and curriculum, teachers and students, then finally students and curriculum. To construct a bridge between the students and the curriculum I will first invite my students to the text. Simply put, I will tell them a story. Not just a story for the sake of a story—a well-told story that will grab the students' interest. The excerpts read to the students should connect to the standards. It is important not to tell the whole story at once. The excerpt or story shared should leave "gaps for students to fill in, holes which encourage them to actively intervene in the proceedings to assume responsibility" (Barone, 2000, p. 62). The gap is the third space. Here in this space, what is said, or sometimes what is not said, will create discourse between students and students, students and teachers, students and the curriculum … just to name a few opportunities. This is the transformative space where students can think about their connection to the story and to the curriculum. Last, the students will need to share their personal reflections with each other. The weaving together of their stories with the stories of others is how the new story, or the new curriculum, will be composed.

I wonder if science itself can become a bridge that can help students find education meaningful, help them become democratic citizens. A bridge where students can linger and make their own connections to science curriculum. Can education become like HeLa cells?

Pequod: How do you propose that education can be like HeLa cells?

Hoffen: Cells can become immortal. Scientists call it immortalization. Immortal "cells do not die out but continue to grow indefinitely if provided with adequate resources" (Van Valen, 1991, p. 71). As a teacher, if I can create a space, complete with adequate resources for students in my classroom to linger and grow and learn continually and indefinitely, could love for learning become immortal?

Think about it. Students respected for who they are and the life experiences they bring to the classroom, building bridges with each other, their teacher, and their curriculum to create meaning ... meaning custom-made to meet their needs. Discourse levels the playing field. Students that have performed poorly in school are able to participate and enter the conversation.

Bradley: In the classroom discourse and "students' personal experiences with the subject matter were explicitly encouraged by the teacher and leveraged upon to delve deeper into the science content at hand" (Basu et al., 2011, p. 58). This allows "deliberate connections [to be] made between school science and the relevant community knowledge students bring with them into the classroom" (p. 58). The story of Henrietta Lacks allows teachers to create this experience for my students. The activity brings the voices of students "into science classrooms in ways that foster critical democracy—not only in how teachers and students enact classroom life together, but also in how students and teachers are supported in leveraging their school experiences towards building a more just world" (p. 58).

Pequod: Here she goes with democracy in science again. Did you not hear my protest last time?

Bradley: Building these bridges and working to achieve a democratic classroom is rather complicated. A basic democratic classroom should include student participation in decision-making that "include curricular scope and focus, classroom participation structures, and rewards and punishments" (Basu et al., 2011, p. 5). I am not talking about simply letting students weigh in on how things happen in the classroom. I am talking about the social and cultural structures that maintain relations of power among students and teachers" (p. 5). In a democratic classroom "the responsibility for shared power and the protection of marginalized voices and perspectives is also elemental" (p. 6). By fostering what a real democracy is in the classroom ... the discourse has a place to happen. Students get more involved with their learning.

Driscoll: I want to go back to Humboldt just one moment. What I take away from your story about Alexander Von Humboldt is the mention of the word discourse. Discourse related to curriculum involves "community politics, bureaucratic regulations, and publishing agendas" (Grumet, 1989/1999a, p. 235). Participants must "determine the rules of that conversation, undermining its promise of open inquiry and democratic participation" (p. 235). Does this sound like a description of a classroom? Or does it sound like life? Remember we discussed that we must bring method into our work with students at any level.

Bradley: In an earlier comment I said it is going to take a "complex, pluralistic, unpredictable system or network" upon which to build a foundation for educational experiences for students and teachers alike (Doll, 1993, p. 3). Sharing the story, then also the subsequent stories composed by your students, does indeed form many intersections. I imagine this approach would certainly get everyone in the classroom involved.

The group hears a knock on the door. Ms. Sonnenschein, the receptionist, opens the door and addresses the group.

Sonnenschein: I hate to interrupt, but I have a tiny little message for Mrs. Hoffen.

Hoffen walks to the door and joins Ms. Sonnenschein outside the door.

Hoffen: Please hurry Krystal … what is it?

Sonnenshein: You are needed in the construction room. Mr. Cooper said it was very important.

Hoffen: Okay Krystal, we will go in just a minute.

Hoffen returns to the meeting.

Bradley: I "use the ideal of democratic education in science to call attention to ways of being in the classroom that position youth as important and powerful participants in their own learning and that of their peers and teachers" (Basu et al., 2011, p. 8). I really like what was said earlier—you bring the world into your method. Now the students see that they are "also members of a larger global society who can leverage their lives in schools towards making a change" (pp. 8–9). The learning in democratic classrooms "position learning as a dialectic process where students and teachers learn to read the word and the world" (p. 7).

Hoffen: Students regardless of academic ability seem to respond well to the story of Henrietta Lacks, and the other stories I share in biology class. It is so sad that our students are assigned a personal value based on their latest benchmark, or standardized test score. Tom Barone (2000) suggests that we need strong poets—"students who continuously integrate the content of disciplines (the content of life) into a coherent and personally relevant world view" (p. 126). His directive given to teachers is simple yet eloquent—"aspire to empowering students within a democratic school setting to act with a sense of personal integrity, responsibility, and autonomy" (p. 126). I want to help all students realize there is a strong poet deep inside them. Our students are not blank slates. Our students come to us "as a slate on which much is already written and where the learner writes new words and phrases in appropriate spots and rearranges phrases to make room for new ones" (Klassen, 2006, p. 826). Standardized testing requires that students wipe off their slate and give the "information back in essentially the same form in which it was originally presented" (p. 826). Strong poets compose their learning. Standardization encourages that knowledge be returned to the teacher in the same form in which it was dispersed—rote memorization. David Blades calls it "Bulimia" (Blades, 1997, p. 72). Having to equate our students' learning with an eating disorder is a very sad situation.

Pequod: (*She speaks quietly.*) Sad? Those with bulimia will perform well on the test; that is not sad. If your students perform well, you might be able to keep your job.

A voice comes over the intercom: "Mrs. Hoffen please report to the construction classroom."

Hoffen: I was hoping we would not be interrupted again. Would you mind taking a walk with me?

The group walks to the construction classroom.

Act VI Scene II: New Teachers in Construction Class

Mandy Hoffen and her entourage walk into a surprisingly empty construction lab. Usually this area was humming with the sound of power tools. Then she turned the corner and saw the classroom area. Two of her biology students were working at the marker board. These two were talking to a whole class of boys. All the boys in the class were listening very intently to what their classmates were saying. Hoffen stood back, so she could listen without being seen.

Matthew*:* Her cells were different. They were cancer cells; they grew very fast. George Gey figured out how to make the cells grow outside the body—this way scientists could study them.

Anthony: Her cells looked like this. (*He drew the outline of a cell on the marker board and drew in two nuclei.*) Cancer cells grow like crazy. They start going through mitosis repeatedly ... you guys remember PMAT? Prophase, metaphase, anaphase, telophase, the steps of mitosis. We know about mitosis because scientists studied Henrietta's cells.

Matthew: It was sad that Henrietta was in a colored ward. They took her cells without her permission or her family's permission. Now that was just wrong. What is worse is that Hopkins was the only hospital for miles that would treat Black patients.

The bell rang, and the students discovered their teacher and the others standing outside the classroom door.

Matthew: We did not know you were here, Mrs. Hoffen. We taught the whole class. Were you standing here the whole time?

Hoffen*:* I only caught the end of class. Wow. Great picture of mitosis, Anthony! You two are great teachers.

Matthew: Mr. Cooper showed us the PowerPoint you sent to all the teachers. Some of the guys had questions. I volunteered to answer. I like this story, Mrs. Hoffen. I feel bad for her family.

Anthony: I did not know that I really understood mitosis until I started talking in front of them. They really had a lot of questions.

The construction teacher, Wayne Cooper, walks up to the group.

Cooper: I am sorry we interrupted the meeting. I thought it was amazing that these two had spent the class period teaching a class full of ninth-grade boys. I figured you would not believe me unless you saw it with your own eyes.

Hoffen: Thanks so much! You guys have made my day ... But now you two better get to class. We will have time to talk later.

The group makes their way through the crowded hallway, back to the conference room.

Driscoll: That is solid evidence that this story is affecting your students, Mrs. Hoffen.

Hoffen: I am still in shock. They were explaining mitosis and talking about issues of social justice. I need you all to understand that Matthew has a specified learning disability. He will only earn a transitional diploma because he cannot pass the math required by the state. At one time, the child was enrolled in three math classes. Anthony suffers from severe ADHD, and often has trouble completing assignments and working in the classroom environment. Both students excelled in biology. Did hearing this story allow them to have success in what otherwise might have been a very difficult class filled with rote memorization? Did mitosis gain meaning because they read Henrietta's story? Both students connected to the story. Both were engaged in science and the story. Both are teenage boys. They just taught a group of teenage boys.

Driscoll: Students are on the very bottom of the power structure. We must remember, "Every individual brings value to society ... more specifically to classrooms. The radical, committed to human liberation, does not become a prisoner" (Freire, 1970, p. 39). Becoming a prisoner mean giving up. The radical or activist "enters into reality so that knowing it better he or she can better transform it" (Freire, 1970, p. 39). Our students need to know they can bring about change. Your young women running around giving out their red beads and wearing their t-shirts have hopes of creating change. The young men teaching their peers want to create change.

Another knock is heard at the door.

Pequod: I will get that. (*She says this obviously annoyed, and terribly bored with the current conversation.*)

Alexis and Lydia: Mrs. Hoffen, Mrs. Hoffen (*The girls speak together*). Mrs. Glover let us tell her class about our shirts.

Alexis: We told the class that the story was more than just HeLa cells … the people, the real people behind the science are important.

Hoffen: Ladies that is wonderful. I think you had better get back to class.

Driscoll: Your students are obviously excited about what they are learning. William Ayers (2004) tells us that "school is a natural site of hope and struggle—hope hovering around notions of a future, struggle busting out over everything about that future: the direction it should take, the shape it could assume, the meanings it might encompass" (p. 20). Teachers and students must participate for there to be struggle. It is in the struggle that both groups of individuals can become educated.

Hoffen: Students must learn that they are stakeholders in their own learning, and they need to become comfortable in the struggle. The students feel powerful when they are making choices about their learning.

Pequod: In my own classroom, I provide students with the answers they need to learn. All they must do is spit them back out to me. This prepares them for their state test. (*She pauses.*) Is this why they become disengaged from their learning?

Bradley: We must look for changes that depart from the norm. Even small a small change "is revolutionary—it represents some departure from the original" (Deever, 1996, p. 188). We as educators and students cannot wait for our entire broken system to undergo a revolution before we are willing to change. Educators and students seem resistant to change although, "the potential exists. If not now, when? If not now, why?" (Deever, 1996, p. 188).

Driscoll: I like what you said about having to get away from the norm for change to occur. I think we have witnessed a change from the norm. Mrs. Hoffen, I see evidence that you are bridging gaps between yourself and the students in your classroom. I see evidence of bridges between the students and curriculum. This can potentially allow students to become motivated to direct their own learning and become activists for something they feel is a worthwhile cause for change.

Hoffen: We have had a great conversation. (*Looks at her watch.*) I would really like to get back to my classroom.

Pequod: We finally agree on something. I would like to get back to my classroom too.

Hoffen: Would you all like to come and hear some of the student presentations?

Bradley: Yes, that is a great idea.

The group stands and walks towards the door.

Act VI Scene III: Student Responses

The group follows Mrs. Hoffen down the hall and enters the classroom. They walk to the back of the room, taking seats in a row of empty desks. A student is holding a computer printout. His hands are visibly shaking.

Jhaymia: I will pull up your beat. (*She walks to the computer.*)

The music starts.

Kyle: I call this HeLa. (*Speaks with a quiver in his voice.*)

Old cells, still livin, I'm rappin bout science.

HeLa cells got labs and ethics and compliance

Ooo, but there is not an alliance.

John Hopkins took them cells—what an act of defiance

But it don't matter anyway

Look how it changed the game today

So many vaccines she led the way

Yea her cells never went a fray

Acting it out, but this isn't a school play

Ooo, Chick Chick Bang:

(Chorus)

You might just get hit with the HeLa

Sellin her cells away but guess what she ain't the dealer!

She been gone for a while; I really wish I could meet her.

A history in the making a true science leader.

Now let's get back to the facts.

Born in Virginia to be exact.

Her real name is Henrietta Lacks.

Her cells deserve to make stacks.

Her tumor cells so strong not even a rock can make a crack.

She is important to the world like fresh breath to a tic-tac. OOO

Shout out to her fam in Roanoke.

Shout out to Henrietta for helping cure polio

Her Helix so fire is catch you with that okie dokee.

She is such a hero the fam should never go broke.

(Chorus)

You might just get hit with the HeLa

Sellin her cells away but guess what she ain't the dealer!

She been gone for a while; I really wish I could meet her.

A history in the making a true science leader.

Switch it up

You might just get hit with the HeLa. (Wilkinson, 2015)

Mrs. Robinson: Wow Kyle! Great job! I think we have time for one more presentation. Charles, it looks like your card is on the top of the deck. (*Charles shyly walks to the front of the room.*)

Hoffen: Awesome Kyle! I wish I had that on video. Have you ever performed a rap before? I wish we had time for more discussion. (*She looks at Mrs. Robinson and then glances at her watch again.*)

Kyle: Well, uh no. I was kind of scared ... but I really wanted to tell my story with a rap.

Charles: I will be honest, I had never written a poem before, but I liked Deborah's poem in the book. So here goes:

Henrietta Lacks
Struck with cancer you were
62 years ago, may seem like a blur
What happened has caused quite a stir
Regardless, your story will be heard

To doctors you were just another cell
They didn't care about you or hope you would get well
Your spirit is gone Earth was your shell
Now all we have left is your story to tell

In a way, I think you would be pleased
To see all the smiles from the lives that you eased
From cloning, gene mapping, and polio vaccines
You have helped the sick and diseased

You may be gone, but your cells are alive
In millions of test tubes is where you survive
On this day is when you died
Immortal forever-worldwide (Yang, 2014)

Hoffen: Charles, I really like your final line ... Immortal forever worldwide. Tell me about poetry. Is this something you would expect to be doing in biology class?

Charles: Not really, I have never liked poetry in English class, but this seemed important. I thought it was a good way to tell Henrietta's Story.

Robinson: Alexis, you are next.

Alexis springs out of her seat and runs to the front.

Alexis: Remember, Mrs. Hoffen let us work in pairs.

Lydia quietly joins her.

Lydia: I want to talk about what it meant for Henrietta to be a woman in 1951.

Alexis: I am going to talk about what it meant for Henrietta to be a Black woman in 1951 in times of segregation and racism.

The bell rings.

>**Hoffen:** We never have enough time. (*She sighs deeply, glancing sadly at her watch.*) Alexis and Lydia, you two will be first tomorrow.

The students pack up and leave Mrs. Hoffen and her committee to finish their discussion.

>**Hoffen:** If a standardized curriculum is required, so be it. I may be required to teach the standards, but I have a lot of control in how I teach the curriculum, and more importantly, how I teach my students.

>**Driscoll:** Students should be respected and be considered curriculum makers right alongside the teacher. All students should be privy to an education that requires "justice, democracy, and human rights" (Pinar et al., 2008, p. 508). Teachers must teach students, not just a standardized curriculum posted on the concrete block wall of the classroom.

>**Bradley:** Henrietta's story provides opportunities for discourse. Allow teachers to bring difficult subjects, such as racism, to the forefront of discussion. *The Immortal Life of Henrietta Lacks* will keep Henrietta's story alive in the hearts and minds of all who hear. By keeping Henrietta's story alive, students can connect themselves to other stories, stories in which they can become participants or activists.

POSTSCRIPT: UNDERSTANDING THE NOISE

The purpose of this conversation was to discuss the elements of curriculum. I meant for the student interruptions to demonstrate how noise can interrupt curriculum and disrupt binary thinking. The dialogue created is an example of discourse, the most necessary element of a living curriculum. Using the text, *The Immortal Life of Henrietta Lacks*, lends itself to lingering. Lingering leads to resurrecting life in science classrooms, by providing students with opportunities to connect their standardized curriculum to themselves. I begin by asking, how many of you have a special woman (mother, wife, grandmother, aunt, sister, cousin, or friend) in your life that you are close to? Thanks, next question: Have you ever been close to someone who experienced a horrible disease? Last question: Has your life or the life of a loved one been impacted by cancer? By this time everyone in the room has responded at least once ... they notice that we all have a lot in common with Henrietta and her family. I think we will all connect to Henrietta in some way. Sometimes the connections are surprising.

I noticed early on, the first year that I used *The Immortal Life of Henrietta Lacks* during our cell cycle unit, that students found the story very interesting. The discussion was exciting. The questions were thoughtful. The emotions ran high as my students voiced their outrage concerning Henrietta being in a public ward, her cells being taken without her permission, and her family having no idea her cells were still alive. Henrietta and her family seemed invisible to science. A young man in my class, Matthew—you may remember him as the construction class teacher—told me that reading about the Black scientists working in the HeLa factory was the first time he had ever read about African Americans doing science. He said it made him feel good to know that his people helped to make a difference. I was saddened that this was the first example he had seen that he himself could become a scientist. "Science's rise to power supposes such a level of recruitment that soon, all-powerful, it creates a vacuum around itself. Which is the reason for the sudden decline of all the surrounding areas of culture—the humanities, arts, religion, and even the legal system" (Serres & Latour, 1990/1996, p. 87). Science will not be developed fully if the "unusual gifts of race have not thereby been developed" (Woodson, 2005, p. 5). Race being ignored in science adds to making some individuals feel invisible or unimportant in the field of science. The narrator in *The Invisible Man* begins by telling us that "I am invisible; understand, simply because people refuse to see me" (Ellison, 1995, p. 3). All our students in science should see themselves and others like themselves as participants in science. Without conversations concerning equity in science education, science education cannot move forward. What I propose may seem too simple … Teachers must build relationships with their students and create a space where the students can connect themselves to the science curriculum. These connections spark life into an otherwise lifeless curriculum. Once the connections are made, students have an opportunity to see themselves in the center of the curriculum as active participants. Active participants who can become activists and instigate change in our world.

Next, I want you to take some time to think about Matthew. He was the special education student that taught his construction class about HeLa cells and Henrietta Lacks. Matthew is a very nice young man with excellent carpentry skills. I have a stool sitting in the corner of my kitchen that he designed and built in his construction class without anyone's help. He was so proud when he brought it down to my room and presented it to me. Because of his weakness in math, Matthew had a math support class to help him with his regular math class. Then during his remediation period, he was selected to receive tutoring for the Math EOCT. He was not allowed to sign up for the chess club with his favorite construction teacher. Basically, Matthew had to endure three math classes a day. His entire day focused on his documented weaknesses. His IEP (Individual Education Plan) stated his weakness was math problem solving, yet he was required to take the same standardized math tests as all the other students in Math I. His scores would be compared to students not requiring an IEP. He viewed himself as a failure. Matthew's experiences in places and situations other than his math class indicate that he has many gifts and academic strengths. Due to public education policy, Matthew was placed on a transitional diploma track that resulted in him leaving school without a diploma.

As you read this story, did you feel connected to Henrietta? Can you relate to Matthew? It is easy to look away, or maybe not look at all. The shock these images create provides interruptions or "noise" in our thoughts. The visual provided by the text creates a locus between the unusual and us ... a new space has been created. Michel Serres (1991, 1997) would consider the space between the viewer and Henrietta or Matthew a third place ... "sown in time and space" (p. 10). A space of transformation will be found in this space between the two opposing binaries. Some might consider this location a bridge, but not in the sense of a bridge constructed with concrete and steel. This bridge is a "site or clearing in which earth, sky, mortals and divine in their longing to be together belong together ... these are not mere paths for human transit, nor are they mere routes for commerce and trade. They are dwelling places for people" (Aoki, 1996, p. 5). They are locations for lingering.

Present-day education is characterized by "modes of learning that promote passivity and rule-following rather than critical engagement on the part of teachers and students" (Giroux, 1983, p. 158). "Increasingly, classrooms are places in which teachers and students act out the script given to them by someone else, neither teachers nor students ask the questions that matter, and learning is equated with passing a test" (Hursh, 2007, p. 3). These characteristics lead to standardized disrespect for students and teachers on multiple levels. We would be kidding ourselves if we do not acknowledge the oppression in many forms that continues to exist in our schools. The standards posted on the concrete block wall in my classroom fail to produce "an increasingly diversified, complicated, and contested cosmos" (He, 2013, p. 63). Instead, the standards are characterized as a "linear sequential, easily quantifiable ordering system," combined with expectations of "clear beginnings and definite endings," a neat and tidy package for teaching, complete with guidelines for delivery (Doll, 1993, p. 3). The predetermined standards lack dimension, as well as meaning, for teachers and students. As students and educators, we can realize the potential of finding a "complex, pluralistic, unpredictable system or network" upon which to build a foundation for educational experiences for students and teachers alike (p. 3).

When individuals are exposed to spaces of transformation and encouraged to linger there or choose to linger there, new characters evolve, characters that are unrecognizable from the original ones encountered. Compare a self, yourself, known for a lifetime, with an "other," with whom there is no prior experience, except possibly oppression. When the two meet in this newly created space, each can possibly be personally transformed. By taking two things and placing those in juxtaposition to each other, noise can be created. Change can occur.

CHAPTER 7

STUDENTS JOIN THE CONSPIRACY AND WEAR RED TO HONOR HENRIETTA LACKS

PRESCRIPT: THE FIRST CELL-E-BRATION

On October 4, 2011, the Evans High School Multicultural Club and Evans High School biology teachers and students invited the entire staff and student body of Evans High School to celebrate the life of Henrietta Lacks. The event was much like the activity going on in the background of Act VI.

On this first Henrietta Lacks Day, teachers at Evans shared information about Henrietta Lacks with each of their classes. We hoped that everyone would hear the story, find it interesting, and want to read Skloot's book for him or herself. Teachers received information to share with students. Everyone was invited to wear red and honor Henrietta Lacks by keeping her story alive. Staff members, Multicultural Club members, and biology students each wore a sticker depicting the book jacket for *The Immortal Life of Henrietta Lacks*. In order to wear a sticker and red ribbon, one must be willing to share the story of Henrietta Lacks with anyone who asked about her. Random House provided permission for use of the book jacket on the stickers as well as the posters that were used throughout the school. Signs to remind students to "Wear Red in Honor of Henrietta Lacks" were posted around the school. Short excerpts from *The Immortal Life of Henrietta Lacks* were shared on the morning school television broadcast for several days prior to the event. During our lunch periods, music from the era when Henrietta Lacks lived and danced on Saturday nights was played in the school com-

Mandy Hoffen and a Conspiracy to Resurrect Life and Social Justice in Science Curriculum With Henrietta Lacks: A Play, pp. 169–213

mons area. Dance tunes of Billie Holliday, Duke Ellington Orchestra, Count Basie, Glen Miller, and Frank Sinatra were featured. Planning for this event started the very first day I heard about Henrietta Lacks. The students, teachers, custodians, lunchroom ladies, and administrators found Henrietta's story to be as amazing as I did on that day in June when my professor shared it with me. We have made this an annual event. This year's celebration marked the 10th anniversary of this event. Over the years additional events have been added. In 2016 the EHS graphic arts students created T-shirt designs to honor Henrietta Lacks (see a sample in Appendix A, Figure 1). A small number of shirts were produced for teachers and students. In 2019 and 2021, students organized art shows in the school's media center to share parts of Henrietta's story (see sample artwork in Appendix A, Figures 2–14). On October 4 each year, the students and faculty of Evans High School don red clothing and share the story of Henrietta Lacks. I personally paint my toenails bright red to honor Henrietta!

This book has chronicled my personal journey with sharing the story of Henrietta Lacks and the evolution of this inquiry. The journey has proven to be a swim in the middle of a raging river (Serres, 1991/1997). To begin the journey, I stepped off the shore, slowly, and eventually found myself in the middle of the river. I don't see this journey ending, or myself finding the other shore, with the composition of the last sentence.

The journey to find alternative pedagogies for the current standardized curriculum will continue. I want the required curriculum that I teach to grow and become a space where I and the students I am privileged to teach can grow, seek change in the world, and become activists for a more democratic society. The required standardized curriculum must be elevated beyond the standardized requirements. I cannot and will not teach science in isolated factual segments that only last one class period. I hope to entice my students to join a conspiracy that will lead to lifelong learning and an immortal love for learning. The journey has involved many conversations and introduced you to many coconspirators who want to bring about change in education. Starting with philosophers and historians of science and then progressing to university professors seeking to improve teaching and pedagogy provided for me a logical progression. I also believed it to be very important to bring the conversation to an actual school. Theory and practice must meet. But the most important aspect of curriculum and learning is what students carry with them when they leave an educational setting. In this instance, the setting is a high school biology classroom.

ACT VII

Act VII is a continuation of the Wear Red to Honor Henrietta Lacks festivities taking place in the previous chapter. The events will culminate in a program to honor Henrietta Lacks. Act VII will focus on the Wear Red to Honor Henrietta Lacks event and the interactions involving teachers, students, scientists, professors, and of course, Henrietta Lacks. They include individuals highlighted in *The Immortal Life of Henrietta Lacks*. Other characters are represented as composites with fictional names. It is in the intersection of these vital participants that a curriculum is created.

Act VII Scene I: Behind the Scenes

Characters in order of appearance for all scenes: Alexis, Lydia, Henrietta, Matthew, Tony, high school students; Mrs. Delores Pequod, high school teacher; Mrs. Mandy Hoffen, high school teacher/graduate student; Dr. George Gey' research biologist; Mrs. Mary Kubicek, laboratory technician; Deborah Lacks, Henrietta's daughter; Lawrence Lacks, Henrietta's son; school board members; Henrietta rising from the HeLa cells; and Henrietta as the mannequin.

The stage is decorated with a Wear Red to Honor Henrietta Lacks banner. On the left of the stage is a wax statue of Henrietta Lacks.

> Her name was Henrietta Lacks, but scientists know her as HeLa. She was a poor Southern tobacco farmer who worked the same land as her slave ancestors, yet her cells—taken without her knowledge—became one of the most important tools in medicine. The first "immortal" human cells grown in culture, they are still alive today, though she has been dead for more than 60 years. (The Lacks Family, 2010).

On the right is a student-produced copy of Pierre-Philippe Freymond's (2006) BioArt display. The school's art club duplicated this display for the schoolwide Wear Red to Honor Henrietta event. The art display consists of a giant neon sign that says "Henrietta's." This sign is hung above a small table holding a microscope, a beaker of HeLa cells, a small notebook that explains Henrietta's story in writing, and a copy of *The Immortal Life of Henrietta Lacks*. Two students, wearing red T-shirts, are standing beside the display.

Alexis: I hope the artist would not mind that this is not a real neon sign?

Lydia: I don't think so, the artist had the same goal that we have … share the story … honor Henrietta Lacks.

Alexis: I really like seeing art and science sharing the stage. Most of the time poets, artists, and thespians like us get labels like geeks and freaks.

Lydia: Mrs. Hoffen says it will be the poets, the artists, and the thespians that make a real difference in this world. Since she is a science teacher you would expect her to say that science is more important than any of the other subjects. Remember what Mrs. Newton read to us in Lit the other day—"Yet again, lo! The soul above all science" (Whitman, 2007, p. 11). I think Whitman is saying souls are more important than science … the soul, the individual, humanity … must be an important factor when we do or study science.

Pequod: (*Walking quickly, she speaks to herself.*) Looks like the pawn shop meets Biology Lab to me. I don't know who has time for all of this. I have got to write lesson plans today, analyze my benchmark data, and prepare students for their Milestone end of course tests and their final exams. That microscope should be in a cabinet before someone breaks it or steals it. Hey, don't touch that (*she yells*) … you break it you buy it. Don't you two have a class today? Shouldn't you be there? (*She rushes off the stage.*) You have been wandering these halls all day.

Alexis: But it is Wear Red to Honor Henrietta today. It all started when HeLa cells became the first cells to grow outside the body … (*She walks toward the teacher, wanting to explain.*) We set up the exhibit. (*The student stops talking when she notices that the teacher is gone. She smiles and shakes her head and mouths—she has no soul—and turns to the other student.*) Did you look at the HeLa cells?

Lydia: I did, when I was setting up the display. They are so cool. I know it is bad for a cell to have two nuclei—too much genetic information for the cell to handle during mitosis—but it does look interesting when you see it under the scope. "The cytoplasm is like a New York City street buzzing," remember (Skloot, 2010, p. 3).

Two young men come on stage and address the two young women.

Daniel: Hey, I think that is the first book I have read since 5th grade.

Matthew: This is just a great story. It really helped me understand what goes on in Biology class. My teacher explains it. But this story, Henrietta, makes it mean something to me.

Alexis: I want everyone to know her story. I mean anything to get Daniel here studying … oops … Don't worry, I won't ruin your rep ... I promise not to tell anyone you stayed up until midnight writing your poem about Henrietta.

Daniel: (*Blushing.*) This is important. I like it. I will probably never forget how mitosis happens or how cancer cells look. Heck, I even know how different types of mutations can happen in cells: point, frameshift, inversion, deletion substitution....

Matthew: I liked learning about her family. Imagine all that Henrietta's kids went through when she died. Then to find out that their mother's cells were still growing all over the world. I wonder if there are any stories to make math interesting.

Lydia: Duh ... Hypatia, Sophie Germain, Ada Lovelace, Sofia Kovalevskaya, Emmy Noether (Zielinski, 2011).

Daniel: Those are all women. I thought you were going to tell me Newton, or Einstein or guys who came up with string theory and chaos theory.

Lydia: Duh! Women bring perspective to all fields, Daniel ... open your male mind to the feminist perspective.

Matthew: Come on man, get in touch with your feminist side. It just means that you support that women are equal. All people are created equal and should be treated with equality.

Alexis: Mrs. Hoffen would be very proud that you two cavemen have crawled out of your cave into the light ... but today is about Henrietta.

Matthew: I never expected to be discussing feminism or racism in science class. We studied syphilis in health class, but we never discussed the Tuskegee Syphilis experiments. All those men died painful deaths ... then there was Crownsville. Why would they pack so many people who need help under one roof? Today people with those same problems can live with their families and even come to school.

Daniel: Crownsville Hospital for the Negro Insane was overcrowded because of budget issues. It became about money, not people.

Lydia: I have never felt so fortunate to have the privilege of just being able to go to the doctor when I am sick. Poor Elsie was our age when she went to Crownsville. We have people with epilepsy that go to school. They just come to school with us. Crownsville was a horrible place. Thank goodness we have health care. Think about Henrietta, if it had not been for

Johns Hopkins leaving all his money for doctors to set up a public ward, she would not have received any medical care at all.

Daniel: Could we enter a new age of segregation because of the Human Genome project. Imagine being denied health care because you have a bad gene. Science is good ... bad ... all at the same time.

Lydia: Medical ethics got a wakeup call when Henrietta entered the picture. But we still have issues with people being treated fairly by science today. What was the guy's name with the spleen? (*Mrs. Hoffen enters stage left.*) Hey Mrs. Hoffen, what do you think?

Hoffen: (*Rushes through.*) I think it looks fantastic. Henrietta looks so real! (*She does a double take, wondering ... is she real?*) You are all doing a fantastic job. I need to make sure the jazz band knows where to get their lunch. I will be back here for the program to start at 1:00 sharp!

Lydia: What were we talking about?

Daniel: John Moore. They decided that his spleen was not his. Once he signed a release form, science could do anything with his spleen that they wanted. Maybe we should read those forms the doctor asks you to sign.

Alexis: You better read. You better study. Education is powerful. It is all about waking up and smelling the coffee—finding as many connections as possible. (*She pretends to head slap Daniel.*) We are weavers. The ideas are threads ... threads that intersect in a woven fabric, "fabric with brilliant threads throughout" (Krall, 1999, p. 5).

Lydia: Connecting science to history, culture, politics, society, and even philosophy is important.

Matthew: Don't forget people and their stories. (*He pauses and looks into Henrietta's eyes.*) She is very pretty. She was a wife, a mother, a daughter, a cousin, a friend ... and her cells revolutionized medical science. "Henrietta Lacks should not have been on the ward because such a segregated space should never have existed" (Wald, 2012a, pp. 187–188).

Lydia: I agree totally Matthew—"and her family should receive state-of-the-art health care because such care should be a basic entitlement" (Wald, 2012a, pp. 187-188). (*Lydia reaches up* and *touches her hand.*) She is pretty, right down to those red toenails.

Alexis: Are you guys going to paint your toes at the booth in the commons area?

Matthew and Daniel stick out their feet for the girls to see their painted toes. The students laugh.

Matthew: I told you they would not notice … Of course, we painted our toes. We want everyone to know we think this story is important. We are taking a stand!

Daniel: In a very peaceful way!

Lydia: I can just imagine Henrietta and her sisters dancing on that hardwood floor every Saturday night.

Jazz begins to play in the background. The girls grab hands and swing dance. The boys join them. The lights fade.

Act VII Scene II: Special Guests Arrive

Mary Kubicek and Dr. George Gey enter, stopping at the HeLa display first.

Mrs. Mary Kubicek: I am glad we have a few minutes before lunch. Did you see the jazz band setting up? I wonder if these kids know how to swing dance.

Dr. George Gey: I saw the band setting up, I think we are in for some excellent entertainment … these students have done a lot of work.

Kubicek: This mannequin looks so real. She is stunning. I will never forget seeing Henrietta in the morgue, that day you sent me to get the last samples of those horrible tumors. I could only look at her chipped nail polish … The students did a great job … look, George, they even painted her toenails red … that is why they call this "Wear Red to Honor Henrietta." When I saw those nails, all those years ago is when I realized "that all those cells came from a real person" (Skloot, 2010, pp. 90–91).

Gey: I never knew one sample of cells could create so much science, so many wonderful discoveries: Help cure polio, help cure cancer. Then came in vitro fertilization and cloning. The story has also created a lot of controversy. I just wanted to rid the world of cancer. We had to have cells to work with outside the body. Margaret and I were not in science to make money, we just wanted to help people ... all people ... Scientists did not get signed consent at the time for samples.

Kubicek: You are preaching to the choir, George. Remember I was there. The world should have known you had good intentions ... I mean you ... we ... worked in a basement, for goodness sake. Cell lines and cell shakers did not need patents in your mind.

Gey: I hope these kids do not see me as a villain today when I tell my part of the story. I thought I was doing the right thing when I did not release her identity. I kept it a secret ... I guess, until Science had to know.

Kubicek: I remember when Stanley Gartler dropped that "HeLa Bomb" at the Chicago meeting (Skloot, 2010, p. 152). I could not believe the genetic tests of those 18 cell lines myself. They should have all been different cells, different donors. I know that shocked everyone who had worked with Henrietta's cells. Funny, I have a hard time calling them HeLa cells now. I can't believe how Gartler described her—she was a cancer victim. Her cells were in the lab. The cells caused the problems. Those cells could travel on dust particles ... how is that any fault of Henrietta's? As a lab technician, I say it was all due to sloppy technique and carelessness. I don't think it is right to personify what the cells are doing onto the donor. She was a real person who should be respected. Henrietta, not just HeLa. She was a real person. Why would you speak ill of a woman you never met? A woman who was raising five children, only to be struck down by cancer. A vicious cancer. She was a victim of her own cells. I suggest keep the two separate. Or can we separate them? The cells and the woman. (*Kubicek looks confused.*) Maybe Gartler would have been more compassionate if he had seen her lying on that table, with all those tumors ... I am ashamed to say I had to look away. That was when I noticed her toes.

Gey: Years of research were ruined. I am sure Gartler did not want to be the one to report the problem. I am sure he thought they would laugh him off the stage.

Kubicek quickly cuts him off.

Kubicek: Reporting the problem was not the issue … how he racialized the whole story … calling the cells promiscuous. The story of a Black woman's cells was told using the racially charged, stereotypical language of the time. Think about it. Would we have HeLa cells growing in culture all over the world if I had not been such a conscientious technician? Think about DNA. Without a woman—Rosalind Franklin—Watson and Crick would still be moving around the pieces of that DNA model. Some scientists would still be spraying people with DDT to keep them from getting polio, if it were not for Henrietta and her cells (Carson, 1962).

Gey: I see your point. Nobody wanted to unleash that story to those researchers that day in Chicago. But today we celebrate. I even wore my red tie. I was glad the Lacks family was able to keep Henrietta's genome private. I am glad that scholarship money has been made available to Henrietta's children. We have made attempts to move forward … but honestly, I think we have a long way to go. I guess if I could go back, I would give more thought to her family. I thought privacy was best.

Matthew approaches Kubicek and Dr. Gey.

Matthew: Hello. Are you Kubicek and Dr. Gey? Mrs. Hoffen sent me to find you. She would like you to come find your seat for lunch.

Kubicek: Yes, we are. You students have done a fabulous job here. I really like the BioArt display. It is nice to see Henrietta honored.

Matthew: Hmm. So, you're the ones that took the cells, kept the cells a secret.

Gey: I did take Henrietta's cells. Mary here grew them. I know now that Henrietta's family should have been included more in the process. That mannequin looks very real.

Matthew: I know. I have this feeling that Henrietta Lacks is right here on this stage. I think she would be happy with our plans for the day. We better go find our places. I think I hear the music.

Act VII Scene III: Members of Henrietta's Family Arrive

Deborah and Lawrence Lacks walk slowly onto the stage.

Deborah Lacks: Oh, Mama! (*She rushes to the mannequin, holding the wax figure's face with her hands and acting very excited.*) Lawrence, doesn't she look real? Don't you think I have her eyes?

Lawrence Lacks: She is beautiful. I cannot imagine how painful her cancer was. I remember everything about her, her face, how she smiled. She smiled most of the time, unless I was in trouble. I do know she probably would not have complained. She was always helping people. She would be happy that her cells helped. I am glad that now we can decide who can use her cells and who can't. She would not want her cells used for the wrong reason ... and they were her cells.

Deborah Lacks: I wish I could remember her. Let's look at the cells. (*They walk across to the microscope.*) Lawrence, I can remember how excited I was when Rebecca took me to the lab to see the cells. Dr. Christoph was the person who really cared enough to tell us the whole story. I did not know what it meant when they said they had my mama's cells in a laboratory. Why didn't they tell us?

Daniel and Matthew enter.

Daniel: Would you like me to show you how to focus the scope?

Deborah peers into the scope, while Daniel narrates how she should focus it.

Deborah: There they are!

Lawrence Lacks: I always thought it was just about the money ... somebody trying to get rich. I am not sure. Like Rebecca Skloot told us and her readers, Science did not have to ask for permission to take samples at the time. Unfortunately, it was a time when racism was rampant ... public wards, those patients in the public ward being used for experiments. Can you imagine if all those rich people who could afford to get the polio vaccine before it was given to everyone would have known that their cure came about because of a Black woman in the colored ward of Johns Hopkins Hospital? You know mama could find good in anything. She would know all of that was wrong, but she would be glad that George Gey decided to let a group of African American scientists take care of Henrietta's cells. This decision paved the way for African American scientists at the Tuskegee institute to become a voice in science. Although I find it ironic that the Tuskegee Syphilis experiments were going on down the hall.

Matthew: I got excited when I learned that African American scientists played an important role in science at the "HeLa factory" (Skloot, 2010, p. 93). We learn about famous scientists, but not many of them look like me. I shared this with Mrs. Hoffen. She sent me to a website so I could see a list. Science is evolving! The list has scientists from all cultures, and many women scientists. Science today is not just White men wearing white lab coats like it was when Henrietta was alive. Science is people, men and women, working together in many different fields … it's … what is that word I am thinking of?

Daniel: Interdisciplinary. Science is not just science. Science is history, philosophy, art, literature, humanities, politics, culture … and more!

Alexis enters stage right and walks to the table holding the microscope.

Deborah Lacks: I hope that what happened to us never happens to another family.

Alexis: Hi, I am so excited to meet you Mrs. Lacks, Mr. Lacks. Mrs. Hoffen asked me to come find you. Lunch is ready! Guys, can you walk our guests to lunch? And then come back to help me with the chairs for the school board and all the other dignitaries … then we will be ready.

The four leave the stage.

Alexis: Where are those chairs? Henrietta, you stay right there. I will be right back to help you move right over there … stage left.

Alexis goes backstage and returns with two chairs, to find the mannequin in a new place.

Alexis: Lydia? Daniel? Matthew? Where are you guys? You guys … How did you get there, Henrietta? One of the guys must have moved you. I better get to lunch.

Lights fade.

Act VII Scene IV: The Program Begins

It is 1:00 P.M. and the auditorium is starting to fill up. Teachers are bringing in groups of students to sit in designated areas. Several students are helping the special guests find their seats beside the podium. Mrs. Hoffen walks to the podium on the side of the stage opposite the art display. Henrietta stands to the left of the podium. Everyone is in place. The sign in the art exhibit is glowing over on the right of the stage. The audience hears a crash of cymbals. A spotlight picks up a young man entering the back of the auditorium. A hip-hop beat fills the auditorium. The young man begins to rap.

Benji:

There once was a man named George Gey

He started doing science like Bill Nye

He was stealing cells from people on the sly

He could've asked but I guess he was too shy

Stealing cells? He's up to the task

Opens up the door and puts on a doctor's mask

Took the cells, then it's back to the lab

Making money off HeLa, yeah, he's rolling fat

HeLa was on Earth but gone too fast

Looked on her different because she was Black

Don't believe me, well it's a matter of the fact

(hook)

Now let me tell you about a woman named Henrietta Lacks

Got her cells on file, we got em by the stacks

So many cells we should start stacking racks

So, listen to this story

Because it's about her past

Her story is as simple as can be

She was an average person like you and me

Went to the doctor and paid a fee

What he told her, she could not believe

He said: you got cancer and it's at stage 3

But then he said let me run a test

See this is the part that I detest

He took her cells and you know the rest

Then he took credit like he was the best

Maybe he simply wanted to protect

Now let me tell you about a woman named Henrietta Lacks

Got her cells on file, we got em by the stacks

So many cells we should start stacking racks

So, listen to this story

Because it's about her past! (Gates, 2016)

The audience breaks out in applause and whistles. Then they hear a loud clap of thunder and see lightning strike the art exhibit on the stage. The neon pink sign flashes. The beaker tips over. The audience hears the glass break and a strange bubbling sound. As the light spreads, the audience can see a shadowy figure, a student whose identity is hidden, kneeling with arms wrapped around the knees. There is a gasp. A spotlight also shines on the shadowy figure where the lightning struck, and the glass broke. The applause ends and the spotlight shifts to Alexis, standing on the stage behind the podium.

Alexis: Thanks, Benji, for sharing Henrietta's story with your rap. A lightning bolt, a flashing sign, and a figure growing up from the cells. Our art exhibit is just one way that we will tell the story of Henrietta Lacks today. The display includes her famous cells, and a picture and booklet highlighting her life. Our work here was inspired by the artist Philippe Freymond (2006). This BioArt exhibit was Freymond's way of telling Henrietta's story. He wanted to show Henrietta's cells and Henrietta as a person together in one piece of art. You will notice that Henrietta has come to life from her HeLa cells. We want you to notice that Henrietta is being resurrected from her HeLa cells today. Keep an eye on Henrietta as we tell her story. In Mrs. Hoffen's class over the last few weeks we have been sharing Henrietta's story through a medium of our choice: art, poetry, letters, science fiction, songs, and many others. We hope that by sharing the story of Henrietta Lacks with all who are here we can keep her story alive. We hope that we can show how stories can be important to learning science.

Hoffen: Thank you, Alexis. Welcome to our Wear Red to Honor Henrietta Lacks event. Today is a day for storytelling. It is also a day when we take a serious look at our science curriculum and ask an important question—Can Henrietta Lacks become part of this high school science curriculum? Welcome to everyone who has found time to attend this cell-e-bration of Henrietta's life. First, I want to introduce Henrietta's daughter, Deborah Lacks.

Deborah Lacks: Thanks so much for inviting me. It is such an honor to be here.

Hoffen: Next, we have Henrietta's son, Lawrence Lacks

Lawrence Lacks: Thanks so much for inviting us, Mrs. Hoffen. As my sister said, it is a real honor to be here. I am so glad you all are sharing our mother's story.

Hoffen: Our visiting scientists today are Dr. George Gey and his lab technician, Mary Kubicek. It was in their lab that HeLa cells began to grow for the first time

Both Dr. Gey and Kubicek smile and wave to the audience. Mrs. Hoffen continues.

Hoffen: I also want to welcome our distinguished school superintendent and our current school board members. (*The members smile and wave.*) You all have a program. We are going to begin today by telling you a story. If you would all just come to the podium when it is your turn. (*The students on the back row stand up and form a line behind the podium.*)

Matthew: "She was a poor Southern tobacco farmer who worked the same land as her slave ancestors, yet her cells—taken without her knowledge—became one of the most important tools in medicine" (The Lacks Family, 2010).

Lydia: On January 29, 1951, Henrietta went to Johns Hopkins Hospital. "I got a knot on my womb," she told the receptionist. "The doctors need to have a look" (Skloot, 2010, p. 13). The doctor found the tumor exactly where she told him he would.

Daniel: A sample of this tumor was taken to Dr. Gey's lab in the basement of Johns Hopkins Hospital. Mrs. Mary Kubicek

(*Daniel and Kubicek exchange a smile*), Dr. Gey's lab technician, fresh out of college, handled all the tissue samples that came into the lab. Kubicek labeled Henrietta's sample—He, the first two letters of her first name, and La, the first two letters of her last name.

Alexis: When looking at Henrietta's records, it was noted that three months prior to this appointment, no tumor had been detected following a six week return visit after the birth of her fifth child. The doctor noted that either they had missed the tumor … or it was growing at an incredible rate.

Daniel: After 2 days, Kubicek was shocked to find that cells were growing. She was even more shocked at the rate they were growing. HeLa cells were growing 20 times faster than normal cells. Mary Kubicek had her hands full taking care of all these cells. Dr. George Gey decided to share with anyone who wanted to work with the cells.

Matthew: After a painful bout with cancer, Henrietta Lacks passed away at the Johns Hopkins Hospital on October 4, 1951.

Lydia: Dr. Gey's lab assistant was sent to the morgue where Henrietta's autopsy was taking place to get a "final sample of cells" (Skloot, 2010, p. 90). (*Lydia smiles and gestures for Kubicek to move up to the microphone.*)

Kubicek: [I] wanted to run out of the morgue and back to the lab, but instead, [I] stared at Henrietta's arms and legs—anything to avoid looking into her lifeless eyes. Then my gaze fell on Henrietta's feet, and [I] gasped: Henrietta's toenails were covered in chipped bright red polish" (Skloot, 2010, p. 90). Mary continued:

> When I saw those toenails, I thought, oh jeeze, she [is] a real person. [I] started imagining her sitting and painting those toenails, and it hit me for the first time that those cells we'd been working with all this time and sending all over the world, they came from a live woman. [I'd] never thought of it that way. (pp. 90-91)

I could not imagine how much she had suffered. She was on that cold table, her body full of tumors. She was just another statistic who died in the public ward. Here I was a White woman, staring at her lifeless Black body. A body deemed less than human by society. She could not just be a woman. She was a Black woman in the public ward. As I stood there, I became very aware of my personal White privilege. I was able to get a college education. I had a job in a major research facility. If I had been sick, I would have been treated in the private patient ward with no expense spared. She was a real woman. Not just HeLa. Not another Black woman in the public ward. She was Henrietta Lacks. I must wonder how many of us would be alive without her. I will let Dr. Gey tell that part of the story.

Dr. Gey: I think her story has touched everyone that hears it. Over the next few weeks, we were amazed that the cells were still living. We were afraid they would stop. No one wants to be embarrassed by their experiments. I went on television to make the announcement. Everything happened so fast! The rest is history. History that I think everyone in a science class should study. HeLa cells led us to a polio vaccine.

Alexis: This is a story about science and how it isn't always what we might think it is. Our teacher began our introduction to the story by reading this passage. Since this is where the story started for me, I want to read it to you.

> Under the microscope, a cell looks a lot like a fried egg: It has a white (the *cytoplasm*) that's full of water and proteins to keep it fed, and a yolk (the *nucleus*) that holds all the genetic information that makes you. The cytoplasm buzzes like a New York City street. It is crammed full of molecules and vessels endlessly shuttling enzymes and sugars from one part of the cell to another by pumping water, nutrients, and oxygen in and out of the cell. All the while, little cytoplasmic factories work 24/7, cranking out sugars, fats, proteins and energy to keep the whole thing running and feed the nucleus—the brains of the operation. Inside every nucleus within each cell in your body, there's an identical copy of your entire genome. That genome tells cells when to grow and divide and makes sure they do their jobs, whether that's controlling your heartbeat or helping your brain understand the words on this page. (Skloot, 2010, p. 3)

Mrs. Hoffen told us about the cytoplasm being as "busy as a New York City street" (Skloot, 2010, p. 3). We thought we would just be seeing parts of the cell defined in a novel. She would tell us stories and we would learn about science. I had always thought science was just about things like the laws of motion, mitochondria, and body systems … atoms.

As the story of Henrietta Lacks is being relayed to the audience, it appears that the figure portraying Henrietta is rising from the spilt HeLa cells. Henrietta is on her knees now, with her arms stretching over her head. Like someone waking up in the morning, taking in deep breaths.

Daniel: "All it takes is one small mistake anywhere in the division process for cells to start growing out of control., just one enzyme misfiring, just one wrong protein" (Skloot, 2010, p. 3). Pro-tee-in. *(he smiles at Alexis)*. I was able to teach my construction class what a cancer cell looks like compared to a normal cell.

A board member looking disgruntled interrupts Daniel.

Board Member 1: Daniel, what is the difference between them?

Daniel: A normal cell has one nucleus. Think about an egg you would cook for breakfast. Most of the eggs, if not all the eggs, you have cracked in a lifetime have one yolk. Normal cells have one nucleus, like how most eggs have one yolk, but a cancer cell can have multiple nuclei—like an egg with two yolks.

Board Member 1: Mrs. Hoffen, our county has a very strict curriculum map. Where in the map does this work with Henrietta Lacks fall? Unless I am mistaken, I do not see her name in the standards for your course.

Henrietta's arms fall; she begins to support her weight with her hands.

Hoffen: Our unit on cells is a huge part of the curriculum. I teach my students about Henrietta Lacks while they are learning about cells. This learning can convey what's covered by the standards but goes beyond that as well.

Henrietta's arms stretch above her again; she begins to breathe deeply.

Board Member 2: Shouldn't your students be doing more activities to prepare them for their Biology Benchmarks? Their End of Course Test?

Henrietta's arms fall; she begins to support her weight with her hands.

Board Member 3: Science has nothing to do with justice, democracy, or human rights. Science is the pursuit of truth and

understanding through a trial of experimentation and verifiable and repeatable results. Do your students carry out experiments with HeLa cells, Mrs. Hoffen?

Alexis: (*very agitated*) Didn't you listen to us? Science is so much
more than just experiments. Don't get me wrong. Experiments
are important. I know how important it was for scientists to
experiment with HeLa cells to find vaccines, and cure cancer.
But science is history, art, poetry, philosophy, culture, stories.
Henrietta and her story are an important part. How can we
learn about HeLa cells and not learn about Henrietta?

*Henrietta's arms stretch above her again, she begins to breathe deeply and rise to
one knee, like she is going to stand up.*

Board Member 3. Science is a matter of experimenting to collect
new data and make knew knowledge. How many formal lab
reports do you have your students write? Do your students
know each step of the Krebs cycle, can they draw all their
amino acids, and can they even draw a simple atom? Do they
know their R-groups?

Matthew: "The Universe is made of stories not of atoms" (Rukeyser,
1968, p. 111).

Henrietta's back arches, she pushes her foot under her and begins to stand up.

Alexis: "We must not see any person as an abstraction. Instead, we
must see in every person a universe with its own secrets, with
its own treasures, with its own sources of anguish, and with
some measure of triumph" (Wiesel, 1992, p. ix).

Lydia: By studying the story of Henrietta Lacks, we can learn about
cells in a different way. We still learn the basics of science, but it
is just more interesting. I know we will all do great on our test.

The restless board member looks as if he is about to speak again.

Hoffen: Thank you, ladies and gentlemen. Sir (*she says sternly*),
"Standards prescribed by governing bodies may serve as
guides and represent expectations, they do not reflect a com-
prehensive understanding of individualized needs and certainly do not account for what happens in classrooms as
children problem-solve and interact with their learning"
(Schultz, 2008, p. 11). And furthermore: "Knowing is a human

way to seek relationship and in the process, to have encounters and exchanges that will inevitably alter us. At its deepest reaches, knowing is always community" (Palmer, 1998, p. 55). Think about how your body is constructed. Do you remember the term connective tissues? Knowledge must be connected to something else. You could use the word build, construct, connect...all would apply. Students can remember what they learn about cells because the information from the standardized curriculum is connected to something else. In this case a story. A story about a real woman, whose cells changed the world, the course of modern medicine as we know it today. (*The board member sighs and decides that he will not continue his current line of argument.*) *Daniel*, why don't you read your poem?

Daniel: *Begins to speak nervously.*

The Tale of Henrietta Lacks

Little girl born in Virginia,

Just a little bit north of Carolina,

She grew up as a nobody

She died a nobody

But was later found as an everybody

Being diagnosed with cancer,

She donated cells without asking why

She died with a silent cry.

However sad, she became our answer,

Mankind's solution

To all the diseases and illnesses,

Henrietta Lacks is our evolution.

She made vaccines and medicine possible

She made life plausible.

HeLa, the immortal cell line,

Population still in the incline.

HeLa, the nobody,

Became HeLa the everybody,

That HeLa saved mankind

Now we see that she is not just HeLa

She is Henrietta Lacks

Forever and ever etched in my mind. (Hall, 2014)

The audience applauds. Henrietta is standing, ready to take a step.

Deborah: Can I read my poem now?

Hoffen: Sure Deborah. This would be a great time. Ladies and gentlemen, please welcome Deborah Lacks, Henrietta's daughter.

Deborah Lacks: (*Looks at her notes, and then looks out to the audience.*) *She begins slowly.*

cancer

checkup

can't afford

White and rich get it

my mother was Black

Black poor people don' have the money to

   pay for it

mad yes, I am mad

We were used by taking our blood and lied to

We had to pay for our own medicine can you believe that?

Johns Hopkin Hospital and all other places,

That has my mother cells, don't give her

Nothing. (Skloot, 2010, p. 280)

My name is Deborah Lacks and I am Henrietta's youngest daughter. It is important that students learn all they can. They don't need to just learn from textbooks. Sometimes textbooks are hard to understand. They need someone to teach them and show them why all that textbook stuff is important. Dr. Christoph at Johns Hopkins cared enough to show my family our mother's cells. He explained what a cell was. He let us look at the cells through a microscope. I thought they were so beautiful. Just like her. To me she was not just HeLa. She was my mother: Henrietta Lacks. It sure was a shock to find out her cells were living after all those years and to find out that "[My] mother was on the moon, she been in nuclear bombs and made that polio vaccine. I really don't know how she did all that, but I guess I'm glad she did, cause that mean she helpin lots of people. I think she would like that." (Skloot, 2010, p. 9)

Henrietta is standing now. Music begins. Lydia stands at the podium and begins to sing to the tune of Owl City's Fireflies. *Henrietta dances to the music on the dimly lit stage.*

Lydia:

Solo

>You would not believe the life,
>
>For it was painful and caused some strife,
>
>Of a lady named Henrietta Lacks
>
>Cervical cancer took over her life,
>
>And because she was a Black woman, wife
>
>She couldn't get the care she needed.
>
>Chorus
>
>I want everyone to see ...
>
>That she matters to me
>
>It is hard to say if we could have made the polio vaccine without her, it might've been just a dream
>
>She wasn't what she seemed

Solo

>In the year 1951
>
>African Americans were widely shunned
>
>So, she had to go to Johns Hopkins hospital
>
>They robbed her of her cells
>
>From the tumors that had begun to swell
>
>And grew them in a lab for years ...

Chorus

>I want everyone to see ...
>
>That she matters to me.
>
>It's hard to say if we could have discovered cloning
>
>Without her it might've been just a dream
>
>She wasn't what she seemed

The scientists got rich …

(Please remember her name …)

But her family didn't get a cent …

(Please remember her name …)

Why do they not tell them?

(Please remember her name …)

That her death was not in vain

Ten million reasons why

Her family spend lots of nights

Crying over their late mom

If they could see her now

She'd make her family so proud

Saving lives not unlike her own

Solo (*slowly*)

I want everyone to see

That she matters to me

If only we'd remember her name

And bring her glory and fame

If only we'd band together and spread, her name …

Henrietta, Henrietta, Henrietta Lacks (Reynolds, 2014)

The music and dancing end and the lights fade.

Act VII Scene V: Curtain Call

Alexis: We want to close with a spoken word. We hope by hearing all the different versions of Henrietta's story today, that you can create your own story … connecting yourself to Henrietta. Remember "everything and everybody has a story" (Baszile, 2008, p. 253). "We bring all our sorted histories, hopes, and desires to the project of curriculum theory, hooking onto familiar stories and creating new ones" (Baszile, 2010, p. 483).

All characters come on stage and make a semicircle. There is a student on each side of the mannequin, and the dancer representing Henrietta remains beside the art exhibit. The students begin to chant.

Standards posted on a brick wall, Rubrics, lesson plans, curriculum maps,

Teachers perform on cue, scripted, prescribed, outcomes predetermined,

Students bubble answers to questions that matter little—Isolated facts, tiny memorized details

Curriculum in dire need of restructuring, revisioning, resurrection

Characterized not by a list of lifeless facts but ideas of Personal, Cultural, Social, Global relevance

Personal Passionate Participatory for all

Pushing Boundaries Crossing Borders

"Inspiring nerves and skin to learn, to remember,

Engaging in ongoing conversations, acting in our worlds" (Miller, 2005, p. 208)

Science and social justice taught side by side,

Transformative for teacher, student, and society

That is the dream

Reality as we know it insists

No time for stories … No time to go off map, no time to discuss, no time to explore

The blank stares on their faces she deplores

Her heart screams

Day after day instead of growing and going deep—we hover but a moment before moving

VIOLENTLY to the next objective

Empty space between the three … the students the curriculum and me.

Could stories fill the gaps? Could stories create cracks in the standards posted on the concrete wall?

Science Stories, student stories, teacher stories

Stories told in the classroom stories told in the halls

Parts of the cell we know it well—we even know about HeLa cells—

He for Henrietta La for Lacks … no one knew where the cells came from …

The real woman fell through the cracks—ignored, invisible, nameless, faceless.

My story, their story, all untold, unspoken—Like Henrietta's until now—

Tumor, radiation, Cells robbed

Cancer treatment in return for experimental participation owed

25 year wait for precipitation of a name, of a person, of a real woman with a real family

Should the story remain untold, is there time? Must we speak of her HeLa cells and move on

NO

October 4th don your red shirts to honor Henrietta

Thinking, reading, writing, talking too

Finding out that it matters who

Science without isolation, Science class with transformation

Social Justice takes the stage, standards on the wall start to fade

Students becoming activists, fighting for social justice.

Students discussed the importance of cells; saw fit to talk about the injustice suffered by Henrietta

Suffered by her family.

What if she had not been Black?

What if she had not been a woman?

What if she had not been poor?

Conversations in class—medical ethics, civil rights, racism, sexism, and violations of basic human rights

Students become the voice of Henrietta—Keeping her story alive

Can this be measured by an end of course test?

Teach to the heart and to the soul—of each one

Noting each special face

Creating each a special space

A space where all participants are privy to social justice

A space where bridges are constructed

The cells and their functions, the real woman too

Why can't both be worthy to note,

The real woman, Henrietta, with chipped red polish on her toes

Come sit, converse, as we compose our story verse by verse

Together we write a story—the philosophers, historians, teachers, students, the curriculum,

And of course, Henrietta!

Everyone on the stage takes a bow, including the Henrietta mannequin that is part of the exhibit. Shock fills everyone's faces. The now-animated mannequin takes a step forward and begins speaking; this was not how their ending had been rehearsed.

Lacks:

Still Alive

With open eyes I see, everything being done to me
As I lay unconnected and detached from reality
I feel their piercing sharp needles press against my anatomy
I'm being robbed of my identity
These "miracle cells" are a part of me
"You can't take them away!" I wish I could say,
But I have no choice
Why must I go through so much agony?
It is already hard enough being a Black woman in society
I must stay strong for my family.
They are all that's left of me
I did not consent to this,
So why must I be forced to face this fatality?

I've been kept in the dark, for what seems like an eternity
And still nothing has changed
No one has heard of me
I'm a real human being, not just a cure for disease
At least share my story if you're going to steal a piece of me

Doctors kept me a secret,
While my cells were healing the rest of humanity
Covering up who I was
To patronize their unrighteous acts of thievery
What they did will never be right
However, this new generation can put up a fight
You can keep my story alive,
By simply whispering my name,
I will live on.

My name is Henrietta Lacks, but the world for such a long time only knew me as HeLa. Thank you all for coming here today to tell my story. There are others with stories that need to be shared. Each of you have your own story to write and tell. Together we weave our stories together in hopes of creating a better world. (Holland, 2016)

The stage goes dark, the curtain falls. When the backstage lights come up, the mannequin is nowhere to be found. The audience reacts with a gasp. Everyone in the audience seems to think this is part of the show. They begin to walk to the exit. Cast members are greeted with bouquets of flowers. The students are in shock. Was

Henrietta pretending to be a mannequin? Did we really see her bow? Did she really speak? Will she become part of science? That is up to the audience members themselves. Will they create their own story and include her? If they create their own story including Henrietta, will they share this story? Will becoming an activist to share stories like Henrietta's become embodied at their very core of being? That story is yet to be told.

Act VII Scene VI: A Walk with Henrietta

Mandy Hoffen runs backstage. She looks in the dressing rooms. She calls for Henrietta.

Hoffen: Henrietta. Henrietta. Where are you? Please don't go. Please wait.

Hoffen sees a shadowy figure at the end of the hallway.

Hoffen: Wait. Please talk to me

Henrietta Lacks: I understand what you are trying to do, Mandy. But what is the point? I was just a woman in the public ward. Lucky me … I had wonderful spectacular cancer cells. I am glad they have helped millions.

Hoffen: You don't have to go away. Your story will be kept alive. People are understanding that you are not just HeLa. You must have thought this would be a good place to be. You traveled a long way to get here. Can we just take a walk? Can we talk?

The two walk out of the hallway door and out onto the sidewalk that surrounds the building.

Hoffen: The lightning strike. You rose from the cells. I knew it. This is crazy. I don't know where to start.

Lacks: Why don't you call me HeLa like everyone else? This story cannot have a happy ending. No matter how many times you say … Henrietta is a real person with a real family, it will not change what has happened. Don't you understand that I am not the only person with a story? My family is not the only family that has witnessed a family member's commodification and exploitation for the sake of science.

Hoffen: I know there are others.

Lacks: What are you going to do about the others, plan a program to honor them too?

Hoffen: I know there are other stories. If my students learn one example, it is my hope that they can apply what they have learned to a new situation. If they understand how your cells were taken, they might become open to listen to stories of others.

Lacks: Hope. You seem to have hope for everything. Your students seemed able to look at my story with different perspectives. So much of what happened is now known. I saw George Gey sitting there. He really thought he was doing the right thing at the time. But today, in 2015, we still have people being exploited for the sake of science.

Hoffen: What is it that you want to happen, Henrietta? What would make things right?

Lacks: For starters, I want science to take care of my family. They should have "health coverage" like your professor suggested (Weaver, 2010, p. 22). Second, I want to make sure those crazy scientists do not do anything to hurt anyone with my cells.

Hoffen: Do you know your family has the right to how your cells are being used (Skloot, 2013)? Also, Rebecca Skloot has worked very hard to make sure your family's genetic privacy has been maintained.

Lacks: I was not trying to be sarcastic about you always having hope. You seem to have acknowledged that it is easier for some to have hope. Don't let me or anyone else take your hope away. You must teach your students how to have hope of a better future. That will be difficult in a world plagued with war, violence, poverty, racism, sexism, disease, global warming, climate change … and I could go on.

Hoffen: I know my privilege does make it easier to have hope. I have so many opportunities at my fingertips—education, health insurance, a great job!

Lacks: It will not be easy for you to convince all those students to take advantage of what is being offered to them. Some of them are just hardheaded kids. I know what hardheaded kids look like. I was in the process of raising 5, remember. (*Henrietta chuckles.*) That little girl Alexis running around is a spitfire. She has your hope. She is determined to change the world. Yeah, I spoke with her when I came into the building. She told me all about me! She was very sweet. I bet you have trouble keeping her in her seat. She gave me these beads. (*Henrietta took a string of shiny red Mardi Gras beads from her suit pocket and places them over her head. She reaches up and grasps the beads.*) You are right Mandy. My story will continue to live here. Not everyone will want to hear my story. A few will. It will be through those few that I will continue to live. Thank you.

Hoffen: Don't thank me. Learning about you and your story has changed my life. If your cells can go to the moon and save millions of people from polio, the least I can do is come to work every day and share your story. Thanks for talking to me.

The two continue walking into the shadows. Alexis comes running out the back door of the school. Yelling loudly.

Alexis: Mrs. Hoffen, Mrs. Hoffen. Where are you going? It is time for the cast party.

Hoffen: I was just walking out here.

Alexis: I could not let you miss the cast party. Everyone is waiting on you. You must be the one to cut the HeLa cell cake!

Act VII Scene VII: Chemistry Lab Down the Hall

Suzy was working at her desk on a test review worksheet. She stopped and looked out the window.

Pequod: Suzy, quit looking out the window. You have an assignment to do.

Suzy: But Mrs. Pequod, I have made good grades on all your tests. I promise I will be ready for my end of course assessment. It is not until May. Why can't we go to the Wear Red program?

Pequod: Get busy Suzy! I have the next chapter review ready for you when you finish this one.

Suzy: Mrs. Pequod, I cannot answer this question. The question is bogus; it cannot be answered.

Pequod: Suzy, you must answer the question. I must verify that all students have answered all the questions. Your growth will determine if I get to keep my job or not.

Suzy: I cannot answer the question. She grabs her head with both hands. I am so confused. We are told it is important to think, and then told just to bubble a stupid answer.

A student yells from the other side of the room, asking the teacher to come and help.

Pequod: Suzy, answer the question

Suzy: Don't you need to go help him? (*She bubbles the answer on the worksheet.*)

Pequod: Good job, that is right!

The teacher puts the paper on her desk. Suzy reaches up and grasps the red Mardi Gras beads around her neck.

Suzy: Is anyone important anymore? Bubble this. Bubble that … Henrietta, your cells are an amazing part of science … but so are you! How can I get people to see that?

POSTSCRIPT: FAST FORWARD TO PRESENT DAY

Evans High School hosted its ninth Wear Red to Honor Henrietta Lacks event on October 4, 2019. My doctoral coursework is completed, and my life has forever changed by participating in reading, writing, and conversations concerning Curriculum over the last nine years when I walked into John Weaver's classroom. My classroom has changed too. An oil painting of Henrietta and other artworks inspired by her story hang on my classroom wall. In this portrait she is depicted with long hair sweeping toward her face as it frames a beautiful smile. Colorful double-helical, deoxyribonucleic acid molecules line the edges of the canvas. This work of art gives face to a woman who remained nameless from her death in 1951 to 1974, when her family learned that her cells were living in laboratories all over the world. A high school biology student created this artwork in order to fulfill the requirements of a project-based assessment for Georgia Performance Standard, "SB1a": "Students will analyze the nature of the relationships between structures and functions in living cells" (Georgia Department of Education, 2011). Part "a" of the standard says that students will "explain the role of cell organelles for both prokaryotic and eukaryotic cells, including the cell membrane, in maintaining homeostasis and cell reproduction" (Georgia Department of Education, 2011). I do not see Henrietta's name in the standard. Do you? Why would I even bring up Henrietta Lacks in a science class? A professor of mine commented on a discussion post— don't see what Henrietta Lacks has to do with Science." In my head, I started working to make a case for Henrietta having a place in science. The more I shared the story of Henrietta Lacks … the more I felt she should have a place in my science classroom, maybe all science classrooms. I also knew there were other stories of importance being ignored in high school science classrooms.

As a classroom teacher of 29 years, I have observed that teachers and students are in a constant struggle to find meaning in these predetermined day-to-day activities mandated by standards and curriculum maps. These trends are not unique to my district. Educational researchers write about the problem. Nussbaum (2010) states that, "Socratic active learning and exploration through the arts have been rejected in favor of pedagogy of force-feeding for standardized national examinations" (p. 19). If these trends continue, we will have "suicide of the soul" (Nussbaum, 2010, p. 142). Teaching to the test often replaces activities that one would expect to find in science classrooms. Unfortunately, it appears that "school science lacks the vitality of investigation, discovery, and creative invention that often accompanies science in the making" (Kokkotas et al., 2010, p. 380). The challenge for science teachers in the current age of standardization involves their finding a way to switch the dial from rote memorization to a "process of discovery and intellectual debate" (Weaver, 2001, p. 20).

In the beginning, my thoughts of changing the curriculum involved making efforts to include stories about science while teaching the standardized curriculum. I found that by adding simple stories to my curriculum, students were more inclined to investigate the required science information. I grew very passionate about the stories I was adding to the curriculum—how could I as a responsible teacher leave out historical accounts of how science was and is being made? The required objectives for biology are very straightforward. "Storytelling could be regarded as a suitable strategy to meet these objectives" (Kokkotas et al., 2010, p. 381). Teaching the standards alone would bring about instruction that to my students may seem abstract and disconnected. Adding stories involving the history of science and the human aspect of how the science came about "has the potential to contribute to the humanizing of teaching, to the improvement of the climate in science classrooms, and to the development of positive attitudes toward science learning. In this context, the understanding of science concepts is expected to improve" (Kokkotas et al., 2010, p. 381).

I also began to think that by learning about the actual people doing science, my students might feel more connected to science. By humanizing the science standards with stories, storytelling allows students to be taught "in a manner that respects and cares for their souls as opposed to a rote assembly line approach" (T. R. Berry, 2005, p. 37). Teaching the standards and the story makes for a natural addition of culturally responsive pedagogy. T. R. Berry (2010) suggests that our plight as teachers must be grounded in "theory and practice" (p. 24), as well as by working hand and hand with students, bringing actual concerns of the community to the table. I must take the theories involving culturally responsive pedagogy, building relationships, and alternative pedagogies into account when planning instruction for my classroom, although the curriculum itself is strictly standardized. "Education should help young people live deeply, fully and wisely" (Morris, 2008, p. 129). Students must have room to explore in-between spaces and linger. Spaces where stories such as the story of Henrietta Lacks is shared must be created. The conspiracy to resurrect Henrietta Lacks will continue if we continue to share her story. Will you join the conspiracy?

APPENDIX A: HENRIETTA LACK'S ARTWORK

This appendix consists of 14 pieces of art created by Evans High School students in response to the school's annual Wear Red to Honor Henrietta Lacks event, specifically 2016, 2019, and 2021. Katie Harris, Evans High School graphic arts instructor, photographed these works. The book cover art is a layering of the art in Figure 7.6 by Kioko Mitchell and Figure 7.9 by Ansley Pacheco.

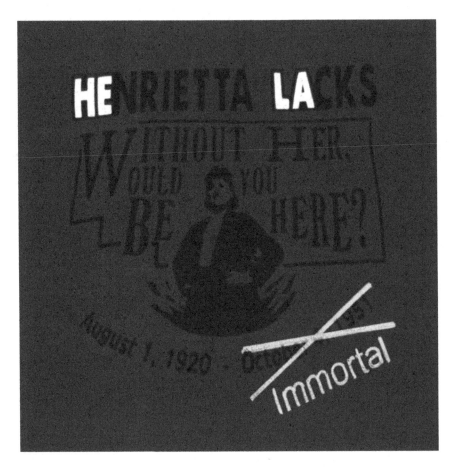

Source: Forbes (2016).

Figure 7.1.   *Henrietta Lacks: Without her would you be here?* [graphic design T-shirt].

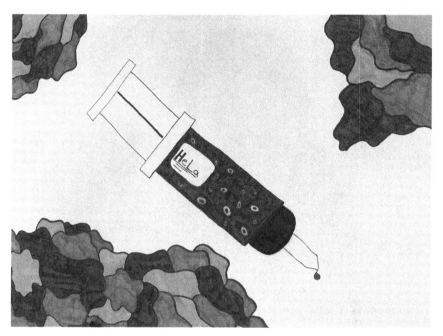

Source: Benefield (2021).

Figure 7.2.   *HeLa cell syringe in color* [marker drawing].

Source: Lan (2021).

Figure 7.3.   *HeLa cells around the modern world* [marker drawing].

Source: M. Lin (2021).

Figure 7.4.   *The story behind the undying cells* [acrylic paint].

Source: V. Lin (2019).

Figure 7.5.   *Beyond HeLa* [pencil drawing].

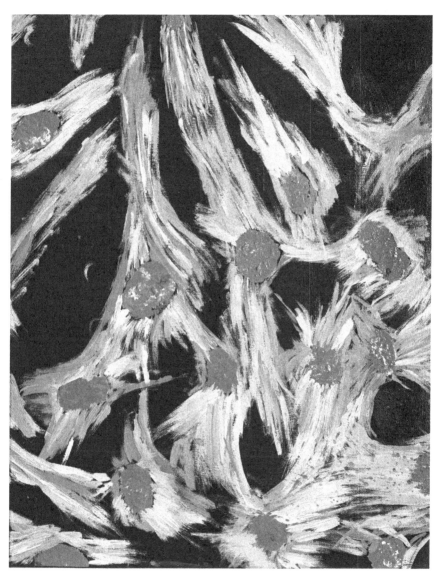

Source: Mitchell (2021).

Figure 7.6.   *Inside Henrietta: More than skin deep* [acrylic paint].

Source: Morton (2021).

Figure 7.7.   *The meaning of Henrietta Lacks* [acrylic paint].

Source: Orr (2021).

Figure 7.8.   *Henrietta in red* [color pencil].

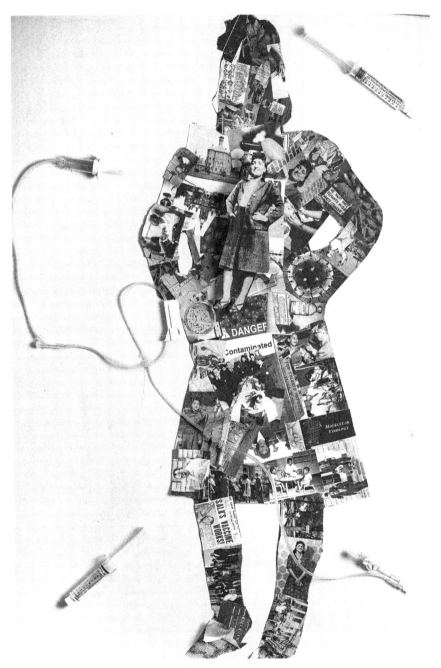

Source: Pacheco (2021).

Figure 7.9. *The inside life of Henrietta* [multimedia collage].

Source: Pruitt (2021).

Figure 7.10.   *Polio's demise* [acrylic paint].

Source: Samuel (2019).

Figure 7.11.   *More than a few cells* [pencil].

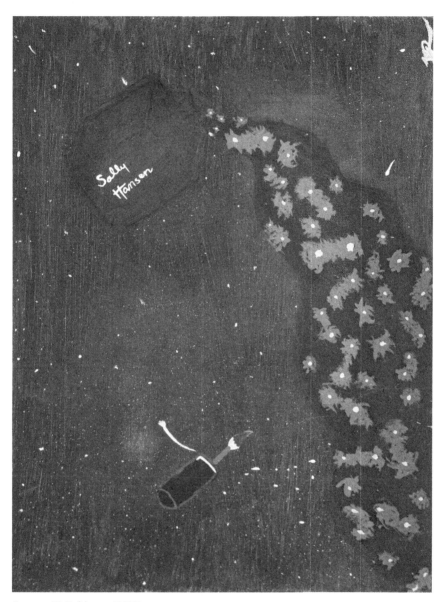

Source: Y. Robinson (2019).

Figure 7.12.   *Red drops of immortality* [acrylic paint].

Source: Simpson (2021).

Figure 7.13.   *The face behind the healing* [pencil drawing].

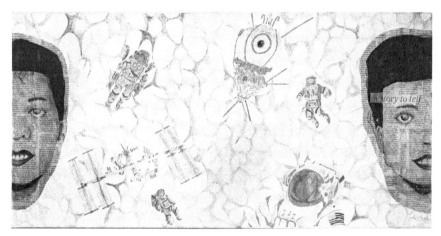

Source: Trimble (2021).

Figure 7.14.   *Never-ending story* [multimedia drawing].

EPILOGUE

Exploring and sharing the story of Henrietta Lacks began with my first reading of *The Immortal Life of Henrietta Lacks*. When I began sharing the story, I was amazed at the conversations the story sparked with my high school biology students. The conversations paved the way for me to connect science to my students who prefer literature, art, and poetry over science. Henrietta's story does indeed take the reader to a place where they can learn "from the recreated other in the text to see features of a social reality that may have gone previously unnoticed" (Barone 1990, p. 314). Conversations concerning scientific research, medical ethics, segregation, racism, sexism, culture, and history have filled my classroom with each reading. Young people have a lot to say. Students need a place where they can linger and develop their voice and speak up for their concerns with the world in which they live. Other opportunities can arise from other stories, but at this moment in time, the story of Henrietta Lacks will remain at the forefront of the curriculum in my classroom. As I read and study, I hope to find other stories that we, my students and I, can explore together.

RESURRECTION OF A LIVING CURRICULUM

Curriculum should be "a coursing, as in an electric current. The work of the curriculum theorist should tap this intense current within, that which courses through the inner person, that which electrifies or gives life to a person's energy source" (Doll, 2000, p. xii). Just as we live with blood coursing through our veins, curriculum is living, breathing, constantly growing, and evolving. Curriculum should incorporate "all the contours

Mandy Hoffen and a Conspiracy to Resurrect Life and Social Justice in Science Curriculum With Henrietta Lacks: A Play, pp. 215–220
Copyright © 2021 by Information Age Publishing
215

of the *Lebenswelt*, the lived world, everyday life" (Bowers, 1995, p. 11). "Curriculum unfolds into the 'curriculum-as-plan' that we typically know as the mandated school subject, and into curricula-as-live(d)—experiences of teachers and students—a multiplicity of curricula as many as there are teachers and students" (Aoki, 2003, p. 2). Where does pedagogy reside? Between the curriculum-as-planned and the live(d) curricula. Aoki calls theses "sites of living pedagogy" (p. 2). The standards on the brick wall in my classroom are in no way living. The standards are a fixed entity in the classroom for teachers and administrators and unfortunately students. If the standards are not taught with elements of lived experience, there will be dire consequences. "Erasing lived experience, erasing human subjectivities in school life, endangers students and teachers alike because we will have no sense of who we are" (Morris, 2001, p. 2).

A curriculum should be interdisciplinary and built from multiple perspectives. A curriculum with potential to resurrect life in science curriculum should incorporate alternative pedagogy that creates a space for curriculum to evolve in spaces where the participants intersect. There must be alternatives for the current one-size-fits-all education implemented by the governing bodies of education. This is not a new dilemma. John Dewey explained during the Progressive education movement over 100 years ago that "The educational experiences must reveal a depth and range of meaning in experiences which otherwise might be mediocre and trivial" (Dewey, 2009, p. 129). Current educational experiences fall very short of the "rich and abundant experience" that education can bring (Grumet 1999b, p. 24).

Connections may deal with a student's interest, a student's background in science, or the value of the student's culture that they bring to the classroom. Curriculum must be designed according to the student's interests and the student's academic needs. A standardized curriculum will never meet the needs of every individual student. As curriculum comes to life, discourse will arise.

CONSPIRACY COMES IN THE FORM
OF DISCOURSE AND LINGERING

Science is discourse. Scientists participate in constant conversations concerning their work in a formal manner at conferences and through peer-reviewed journals. Science education should involve discourse. Participation in discourse enables students and teachers to find a broader understanding of curriculum not only through discussions concerning science, but by seeing connections to "cultural, historical, political, ecological, aesthetic, theological, and autobiographical" aspects of education (Slattery,

2013, p. 200). Teachers and students must also realize the "impact of the curriculum on the human condition, social structures, and the ecosphere rather than the planning, design, implementation, and evaluation of context-free and value-neutral schooling events and inert information" (Slattery, 2013, p. 200). Meaning will rise from discourse. Education should be a meaningful experience for participants. Students' success should not require them to abandon their defining characteristics: "ethnicity, race, and sex, all of which are stripped away" (Pinar et al., 2008, p. 304). Students' personal stories cannot be discounted. Students must be considered valued stakeholders in the education process. I propose that storytelling in science serve as a springboard to create the needed discourse to make science meaningful to students.

Without stories, curriculum can be characterized as a "linear sequential, easily quantifiable ordering system," combined with expectations of "clear beginnings and definite endings," prepackaged for teaching complete with guidelines for delivery (Doll, 1993, p. 3). Blades (1997) uses the term "technical-rational approach" to describe this prepackaged phenomenon. The "technical-rational approaches to curriculum change have dominated educational discourse since the mid-seventies" (p. 84). Nothing about curriculum changes, because models created do not actually change. The models focus on techniques of teaching. "Instead of examining the model itself, the failure of curriculum reform has been rationalized during the past decade as a problem of knowledge: change has yet to succeed because the correct techniques have not been fully articulated and then applied" (p. 84). If the model curriculum were delivered in an appropriate manner by all teachers, the model would be successful in promoting change. If teachers were machines, the proposed models would prove effective. Joel Spring (2013), in his recent satire *Common Core*, tells a story of terrorism in schools. He speaks about how curriculum could become successful by simply changing the delivery method. Fictional educational leaders in his book address the issue by suggesting that replacing teachers with robots would solve the problem of failed curriculum initiatives. Spring (2013) writes: "Teacher robots: Our hope, our future" (p. 48). Teachers often feel that administrators and education initiatives require them to be robotic in order to meet all the evaluation standards.

The vicious cycle of attempts to perfect teaching techniques goes one step further, mandating teachers to use a strictly standardized curriculum. Along with current scripted delivery methods, the predetermined standards lack dimension, as well as meaning, for teachers and students. The potential for improvement would involve finding a "complex, pluralistic, unpredictable system or network" upon which to build a foundation for educational experiences for students and teachers alike (Doll, 1993, p. 3).

Marla Morris (2001, p. 103) describes curriculum as a collection of noise in the classroom; not physical noise such as 32 students talking at once or potential disruptions over the classroom intercom but the "process where we feed off one another, we interrupt taken-for-granted knowing's, we generate new orders out of disorders." This type of living, noisy, curriculum cannot be prescripted nor put on a curriculum map.

Henrietta Lacks has provided a lot of noise in my classroom. A space has evolved for a living curriculum full of discourse. It is in this place that we can understand how the history, culture, and philosophy of science and science education intersect with students. Transformation cannot take place without venturing outside boundaries and crossing predetermined borders. Teachers must venture outside these borders and provide an opportunity for students to venture outside the lines of the usual linear standardized education. The black box is sitting there, waiting. A scientific system inside the black box, left for only scientists, must be opened and made available to all participants of science. These participants can gain understanding and social justice by constructing connections in "spaces of transformation" (Serres, 1982/2007, p. 73). They can also be provided ample opportunities to linger within those connections and spaces.

Lingering, such as in the context of the play presented here, provides an opportunity for us to compose a third narrative of Henrietta, where she is returned to situated "locations" in a Humanity actively aware of racism (Clarke, 2008, p. 2). What must happen in order to enhance our capacity for lingering on our collective belonging as ever-changing humans? How to dwell on the problem of life itself? Education offers the perfect grounds for lingering and turning to the kind of political "mindfulness," Colebrook (2014) suggests, yet current standardized educational practices deny researchers, teachers, and students these opportunities (p. 137). For those who read Skloot's book or see the recently released HBO movie, "The Immortal Life of Henrietta Lacks," an opportunity for lingering is created (Wolfe, 2016). Without the sharing of her story, Henrietta would remain nameless and faceless.

HeLa cells continue to be used in research under the watchful eye of the Lacks family. What will happen to Henrietta? Colebrooke (2014) tells us that "the condition for any being's survival, 'it's living on,' is that it take some distinct repeatable form" (Colebrook, 2014, p. 210). What form will be remembered here? Henrietta and HeLa cells are both living entities. I propose that this repeatable form include both the woman and the cells. Both need to live on in classrooms to broaden students' perspective of racism, science, and the world in which they live.

CLOSING THOUGHTS

I want to teach my students how to linger. With curriculum maps and benchmarking deadlines, it is very difficult to create an atmosphere fit for lingering. We must stop; we must be interrupted for real lingering to occur. Noise must shatter our original thinking—our thought patterns for us to venture into a new space where we can compose new stories. "We bring all our sorted histories, hopes, and desires to the project of curriculum theory, hooking onto familiar stories and creating new ones" (Baszile, 2010, p. 483). My story has been rewritten by knowing the story of Henrietta Lacks. My story has been complicated by the all the authors I have read, real conversations in my graduate classes, real conversations with my high school students, digital conversations, and the conversations that only took place in my imagination. As we read, write, talk, and think, the conversation will continue to be both complicated and rewarding.

As we part, I reflect that I began with intentions of writing a narrative that the reader would find inspiring and simple to follow. The intention of this storied work was to complicate the readers' story or stories. As a participant in this conspiracy you had an opportunity to linger with and connect to Henrietta Lacks. I wonder: Will your journey continue? Will the connections that you made continue to inform your personal conspiracy to work toward change in education? All participants have an opportunity for continued growth as stories are shared and woven together and as other conspiracies arise to instigate change in education. In the process of writing, as a participant in this conspiracy, my story has been complicated. This is not a journey that must end when the last page is turned. I will continue to linger.

Rebecca Skloot provided the results of her 10-year inquiry in a manner for the entire world to hear. The book—rich with lessons in science, racism, history of science, medical ethics, and social justice—is a must-read. A life forgotten is resurrected through the reading of this text and returned to its rightful place in the science classroom. I close with words from a ninth-grade biology student:

> We can take her story and make it our own. We have the power to let her voice be heard! We shall make the ground shake with the heaviness of our hearts, and the sadness of her voice. Henrietta will not be silenced! She shall not be forgotten! She shall be known as the woman that gave life. The woman that cured diseases. The woman that saved the human race from all of the things that tried to destroy us. Henrietta Lacks is more than a woman. Henrietta is a hero. She is our hero. (J, Robinson, 2011)

This student identified Henrietta Lacks as an important part of herself and her education, sustaining the act of lingering between a required standardized curriculum and the lived experience of Henrietta Lacks. Can lingering create similar results in education more generally? Colebrook, tells us "perhaps it is only in our abandonment of ownness, meaning, mindfulness and the world of the body that life, for whatever its worth, has a chance" (Colebrook, 2014, p. 138). Lives are touched, the living are connected to science, and the living narrative is resurrected, becoming part of curriculum and curriculum studies through meaningful stories such as *The Immortal Life of Henrietta Lacks*. May Henrietta Lacks become part of your complicated conversations …

REFERENCES

Adichie, C. (2009, October). *Chimanda Adichie: The danger of a single story.* http://www.ted.com/talks/chimamanda_adichie_the_danger_of_a_single_story/transcript?language=en

Aoki, T. T. (1996). *Imaginaries of "East and West": Slippery curricular signifiers in education.* Paper presented at the International Adult & Continuing Education Conference, Norman, OK.

Aoki, T. (2003). Localizing living pedagogy in teacher research: Five metonymic moments. In E. Hasebe-Ludt & W. Hurren (Eds.), *Curriculum intertext* (pp. 1–10). Peter Lang.

Appelbaum, P. (2001). Pastiche science. In J. A. Weaver, M. Morris, & P. Appelbaum (Eds.), *(Post) modern science (education): Propositions and alternative paths* (pp. 111–127). Peter Lang.

Appelbaum, P., & Clark, S. (2010). Science! fun? A critical analysis of design/content/evaluation. *Journal of Curriculum Studies, 33*(5), 583–600.

Arendt, H. (1951). *The origins of totalitarianism.* Harcourt Brace Jovanovich.

Ayers, W. (2004). *Teaching toward freedom.* Beacon.

Barone, T. (1990). Using the narrative text as an occasion for conspiracy. In E. W. Eisner & A. Peshkin (Eds.), *Qualitative inquiry in education: The continuing debate* (pp. 305–326). Teachers College Press.

Barone, T. (2000). *Aesthetics, politics, and educational inquiry.* Peter Lang.

Barone, T., & Eisner, E. W. (2012). *Arts based research.* SAGE.

Barton, A. C., Ermer, J. L., Burkett, T. A., & Osborne, M. D. (2003). *Teaching science for social justice.* Teachers College Press.

Basu, S. J., Barton, A. C., & Tan, E. (2011). *Democratic science teaching: Building the expertise to empower low-income minority youth in science.* Sense.

Baszile, D. T. (2008). Beyond all reason indeed: The pedagogical promise of critical race testimony. *Race Ethnicity and Education, 11*(3), 251–265.

Baszile, D. T. (2010). In Ellisonian eyes: What is curriculum theory? In E. Malewski (Ed.), *Curriculum studies handbook: The next moment* (pp. 483–495). Routledge.

Mandy Hoffen and a Conspiracy to Resurrect Life and Social Justice in Science Curriculum With Henrietta Lacks: A Play, pp. 221–229

Copyright © 2021 by Information Age Publishing

221

Benefield, K. (2021). *HeLa cell syringe in color* [Marker drawing]. Henrietta Lacks Art Show, Evans High School, Evans, GA.

Bennet-Smith, M. (2012, August 10). *Wang Shangkun, Chinese teen who sold kidney to buy iPad, too weak to face alleged harvesters in trial.* http://www.huffington-post.com/2012/08/10/wang-shangkun-kidney-ipad_n_1764335.html

Berry, B. (2007). *The reauthorization of No Child Left Behind: Views from the nation's best teachers.* Retrieved from Teacher Leaders Network: http://www.teachingquality.org/sites/default/files/TLN%20on%20NCLB.pdf

Berry, T. R. (2005). Black on black education: Personally, engaged pedagogy for/ by African American pre-service teachers. *The Urban Review, 37*(1), 31–48.

Berry, T. R. (2010). Engaged pedagogy and critical race feminism. *Educational Foundations, 24*(3/4), 19–26.

Best Sellers. (2013, June 8). http://www.nytimes.com/best-sellers-books/paperback-nonfiction/list.html

Blades, D. W. (1997). *Procedures of power & curriculum change: Foucault and the quest for possibilities in science education.* Peter Lang.

Blades, D. W. (2001). The simulacra of science education. In J. A. Weaver, M. Morris, & P. Appelbaum (Eds.), *(Post) modern science (education): Propositions and alternative paths* (pp. 57–94). Peter Lange.

Bowers, C. A. (1995). *Educating for an ecologically sustainable culture.* SUNY.

Brody, A. (2011). Operation epsilon: Science, history, and theatrical narrative. *Narrative, 19*(2), 253–257.

Butler, J. (2005). *Giving an account of oneself.* Fordham University Press.

Butler, O. (1987). *Dawn: Book one of the xenogenosis series.* Warner.

Callaway, E. (2013, March 27). *HeLa publication brews bioethical storm.* http://www.nature.com/news/hela-publication-brews-bioethical-storm-1.12689

Carlson, D. (2015). Forward. In N. Snaza & J. A. Weaver (Eds.), *Posthumanism and educational research* (p. x). Routledge.

Carson, R. (1962). *Silent spring.* Houghton Mifflin.

Cartwright, N. (1999). *The dappled world: A study of the boundaries of science.* Cambridge University Press.

Casti, J. L. (1998). *The Cambridge quintet: A work of scientific speculation.* Little, Brown and Company.

Clarke, B. (2008). *Posthuman metamorphosis: Narrative and systems.* Fordham University Press.

Colebrook, C. (2014). *Death of the posthuman: Essays on extinction, Vol. 1.* Michigan.

Coles, R. (1989). *The call of stories: Teaching and the moral imagination.* Houghton Mifflin Company.

Coles, R. (1997). *Doing documentary work.* Oxford Press.

Connelly, M. F., & Clandinin, D. J. (2000). *Narrative inquiry: Experience and story in qualitative research.* Jossey-Bass.

Curtis, A. (Director). (1997). *The way of all flesh* [Motion Picture].

Deboer, G. E. (1991). *A history of ideas in science education: Implications for practice.* Teachers College Press.

Deever, B. (1996). If not now, when? Radical theory and systemic curriculum reform. *The Journal of Curriculum Studies, 28*(2), 171–191.

Derrida, J. (1974). *Of grammatology* (Gayatri Chakravorty Spivak, Trans.). John's Hopkins Press. (Original work published 1967)

Dewey, J. (2009). *Democracy and education.* Feather Trail Press.

Dickinson, E. (1993). *Emily Dickinson.* Everyman's Library; Reprint edition (November 2, 1993).

Doll, M. A. (2000). *Like letters in running waters.* Erlbaum.

Doll, W. E. (1993). *A post-modern perspective on curriculum.* Columbia University.

Doll, W. E., Feng, F., & Petrina, S. (2001). The object(s) of culture: Bruno Latour and the relationship between science and culture. In J. A. Weaver, M. Morris, & P. Appelbaum (Eds.), *(Post) modern science (education): Propositions and alternative paths* (pp. 25–39). Peter Lang.

Duke University. (2015, June 1). Priscilla Wald. http://english.duke.edu/people?subpage=profile&Gurl=/aas/English&Uil=pwald

Egan, K. (1986). *Teaching as storytelling: An alternative approach to teaching and curriculum in the elementary school.* University of Chicago Press.

Einstein, A. (1997, February 20). *Albert Einstein's letters to President Roosevelt, August 2, 1929.* http://hypertextbook.com/eworld/einstein.shtml

Ellison, R. (1995). *The invisible man.* Vintage International.

Emerson, R. W. (2003). *Ralph Waldo Emerson: Nature and Selected Essays.* Penguin Books.

Every Child Succeeds Act of 2015, Public Law No. 114-95, S.1177, 114th Cong. (2015). Retrieved from https://www.congress.gov/114/plaws/publ95/PLAW-114publ95.pdf

Ferguson, S. (2014, July 25). July 20, 2014. Looking forward. *Sara's adventures in education.* http://epiceducation.blogspot.com/

Feyerabend, P. (1975/2010). *Against method.* Quad Graphics.

Forbes, E. (2016). *Henrietta Lacks: Without her would you be here?* [Graphic design T-shirt]. Henrietta Lacks Art Show, Evans High School, Evans, GA.

Freire, P. (1970). *Pedagogy of the oppressed.* Continuum.

Freymond, P.-P. (2006). *HeLa.* (P.-P. Freymond, Performer) Centre Intégratif De Génomique, Université de Lausanne, Dorigny, Lausanne, Switzerland.

Gartler, S. (1967). The genetic markers as tracers in cell culture. *Second Decennial Review: Conference on Cell Tissue and Organ Culture Proceeding, 26*(1), 67–190.

Gates, S. (2015, October 4). HeLa Rap (S. Gates, Performer). Evans High School, Evans, GA.

Gayler, K. (2005). *How have exit exams changed our schools? Some perspectives from Virginia and Maryland.* Center on Education Policy.

Georgia Department of Education. (2011). https://www.georgiastandards.org/Standards/Pages/BrowseStandards/ScienceStandards.aspx

Giroux, H. A. (1983) *Theory and resistance in education: A pedagogy for the opposition.* Bergin & Garvey.

Giroux, H. A. (2009). Obama's dilemma: Post partisan politics and the crisis of American education. *Harvard Educational Review, 79*(2), 250–266.

Gold, M. (1985). *A conspiracy of cells: One woman's immortal legacy and the medical scandal it caused.* SUNY Press.

Grumet, M. R. (1990). On the daffodils that come before the swallow dares. In E. Eisner & A. Peshkin (Eds.), *Qualitative Inquiry in education: The continuing debate* (pp. 101–120). Teachers College Press.

Grumet, M. R. (1999a). Word worlds: The literary reference for curriculum criticism. In W. F. Pinar (Ed.), *In contemporary curriculum discourses: Twenty years of JCT* (pp. 233–245). Peter Lang. (Original work published 1989)

Grumet, M. R. (1999b). Autobiography and reconceptualization. In W. F. Pinar, *Contemporary curriculum discourses: Twenty years of JCT* (pp. 24–29). Peter Lang.

Guattari, F. (1995). *Chaosmosis: An ethico-asesthetic paradigm* (P. Bains & J. Pefanis, Trans.). Indiana University Press.

Hall, H. (2013, October 4). *The tale of Henrietta Lacks* (H. Hall, Performer). Evans High School, Evans, GA.

Haraway, D. (1989). *Primate visions: Gender, race, nature in the world of modern science.* Routledge.

Haraway, D. (1991). *Simians, cyborgs, and women: The reinvention of nature.* Free Association Books.

Haraway, D. (1997). *Modest_witness@second_millennium.femaleman_meets_Onco-Mouse.* Routledge.

Haraway, D. (2000). *How like a leaf: An interview with Thyrza Nichols Goodeve.* Routledge.

Harding, S. (1991). *Whose science? Whose knowledge? Thinking from women's lives.* Cornell University Press.

Harding, S. (1998). *Is science multicultural? Postcolonialisms, feminisms, and epistemologies.* Indiana University Press.

He, M. F. (2003). *A river forever flowing: Cross-cultural lives and identities in the multicultural landscape.* Information Age.

He, M. F. (2013). East~West epistemological convergence of humanism in language, identity, and education: Confucious~Makiguchi~Dewey. *Journal of Language, Identity and Education, 12*(1), 61–70.

He, M. F., & Phillion, J. (2008). *Personal~passionate~participatory inquiry into social justice in education.* Information Age.

Heese, B. (2015). *Effects of the elimination of grade 12 provincial exams in chemistry, biology, and physics on teachers in a British Columbia school district* (Unpublished doctoral dissertation). British Columbia, University of Victoria.

Hodson, D. (1998). *Teaching and learning science: Toward a personalized approach.* Open University Press.

Holland, J. (2015, October 4). *Still alive* (J. Holland, Performer). Evans High School, Evans, GA.

hooks, b. (2010). *Teaching critical thinking: Practical wisdom.* Routledge.

Humboldt, A. v. (2011). *Researches, concerning the institutions and monuments of the ancient inhabitants of America, with descriptions and views of some of the most striking scenes in the Cordilleras* (H. M. Williams, Trans.). Cambridge University Press. (Original work published 1814)

Humboldt, A. v. (1993). *Personal narrative of a journey to the equinoctial regions of the new continent 1799-1804* (J. Wilson, Trans). Penguin Group. (Original work published 1834)

Humboldt, A. v. (1997). *Cosmos: A sketch of the physical description of the universe* (E. C. Otte, Trans). The Johns Hopkins University Press. (Original work published 1858)

Hursh, D. (2007). High-stakes testing and the decline of teaching and learning: The real crisis in education. *American Educational Research Journal, 44*(3), 493–514.

Innes, C. (2002). Bridging opposites: Drama and science in the plays of John Mighton. *Canadian Theater Review,* 20–26.

Jones, H. W., McKusick, V. A., Harper, P. S., & Wuu, K.-D. (1971). After office hours: The HeLa cell and a reprisal of its origin. *Obstetrics and Gynecology, 38*(6), 945–949.

Jupp, J. C. (2004). Culturally relevant teaching: One teacher's journey through theory and practice. *Multicultural Review, 13*(1)33–40.

Klassen, S. (2006). A theoretical framework for contextual science e teaching. *Interchange, 37*(1–2), 31–62.

Kokkotas, P., Rizaki, A., & Malamitsa, K. (2010). Storytelling as a strategy for understanding concepts of electricity and electromagnetism. *Interchange, 41*(4), 379–405.

Krall, F. R. (1999). Living metaphors: The real curriculum in environmental education. In W. F. Pinar (Ed.), *Contemporary curriculum discourses: Twenty years of JCT* (pp. 1–5). Peter Lang.

Kumar, R. (2012, August 28). *An open letter to those colleges and universities that have assigned Rebecca Skloot's the immortal life of Henrietta Lacks as the "common" freshmen reading for the class of 2016.* http://itsbrowntown.blogspot.com/2012/08/an-open-letter-to-those-colleges-and.html

Lacks, L., Lacks, B., & Lively, A. D. (2013). *HeLa family stories: Lawrence and Bobbett (A short memoir).* A Kindle Edition. HeLa Family Enterprises.

Lake, R. (2008). A curriculum of imagination in an era of standardization. In J. Phillion & M. F. He (Eds.), *Personal~passionate~participatory inquiry into social justice in education* (pp. 109–125). Information Age.

Lan, L. (2021). *HeLa cells around the modern world* [Marker/crayon drawing]. Henrietta Lacks Art Show, Evans High School, Evans, GA.

Landecker, H. (1999). Between beneficence and chattel: The human biological in law and science. *Science in Context, 12*(1), 203–225.

Landecker, H. (2000). Immortality, invitro: A history of the HeLa cell line. *Biotechnology and Culture: Bodies, Anxieties, Ethics,* 53–72.

Landecker, H. (2007). *Culturing life: How cells became technologies.* Harvard University Press.

Lather, P. (2007). *Getting lost: Feminist efforts toward a double(d) science.* State University of New York.

Lather, P. (2010). *Engaging science: Policy from the side of the messy.* Peter Lang.

Lather, P., & Smithies, C. (1997). *Troubling the angels.* Westview.

Latour, B. (1987). *Science in action: How to follow scientists and engineers through society.* Cambridge: Harvard University Press.

Latour, B. (1993). *We have never been modern* (C. Porter, Trans). Harvard University Press. (Original work published 1991).

Latour, B. (1999). *Pandora's hope: Essays on the reality of science studies.* Harvard University Press.

Lin, M. (2021). *The story behind the undying cells* [Acrylic paint]. Henrietta Lacks Art Show, Evans High School, Evans, GA.

Lin, V. (2019). *Beyond HeLa* [Pencil drawing]. Henrietta Lacks Art Show, Evans High School, Evans, GA.

Mighton, J. (1987). *Scientific Americans.* Playwrights Canada Press.

Mighton, J. (1988). *Possible worlds.* Playwrights Canada Press.

Miller, J. L. (2005). *Sounds of silence breaking.* Peter Lang.

Mitchell, K. (2021). *Inside Henrietta: More than skin deep* [Acrylic paint]. Henrietta Lacks Art Show, Evans High School, Evans, GA.

Morris, M. (2001). Serres bugs the curriculum. In J. A. Weaver, M. Morris, & P. Applebaum (Eds.), *(Post) modern science (education): Propositions and alternative paths* (pp. 94–110). Peter Lang.

Morris, M. (2008). *Teaching through the ill body: A spiritual and aesthetic approach to pedagogy and illness.* Sense.

Morrison, T. (2008). *A mercy.* Vintage Books.

Morton, J. (2021). *The meaning of Henrietta Lacks* [Acrylic paint]. Henrietta Lacks Art Show, Evans High School, Evans, GA

Nussbaum, M. C. (2010). *Not for profit: Why democracy needs the humanities.* Princeton University Press.

Orr, G. (2021). *Henrietta in red* [Color pencil]. Henrietta Lacks Art Show, Evans High School, Evans, GA.

Pacheco, A. (2021). *The inside life of Henrietta* [Multimedia collage]. Henrietta Lacks Art Show, Evans High School, Evans, GA

Palmer, P. (1998). *The courage to teach.* Josey-Bass.

Philbrick, N. (2011). *Why read Moby-Dick?* Viking Penguin Group.

Pinar, W. F. (1994). *Autobiography, politics and sexuality: Essays in curriculum theory 1972-1992.* Peter Lang.

Pinar, W. F. (2004). *What is curriculum theory?* Erlbaum.

Pinar, W. F., Reynolds, W. M., Slattery, P., & Taubman, P. M. (2008). *Understanding curriculum.* Peter Lang.

Plato. (1976). *Meno.* Hacket. (Original work published 360 B.C.)

Pruitt, B. (2021). *Polio's demise* [Acrylic paint]. Henrietta Lacks Art Show, Evans High School, Evans, GA.

Reynolds, R. (2011, October 4). *Henrietta's dance* (R. Reynolds Performer). Evans High School, Evans GA.

Rheinberger, H. J. (1992). Experiment difference and writing: Tracing protein synthesis. *Studies in History and Philosophy of Science, 23*(2), 305–331.

Rheinberger, H. J. (2010). *An epistemology of the concrete.* Duke University Press.

Robinson, J. (2011, October 4). *The immortal wonderland* (J. Robinson, Performer). Evans High School, Evans, Georgia.

Robinson, Y. (2019). *Red drops of immortality* [Acrylic paint]. Henrietta Lacks Art Show, Evans HighSchool, Evans, GA.

Rogers, M. (1976, March 26). The double-edged helix. *The Rolling Stone, 209,* 48–51.

Rowlands, M. (2009). *The philosopher and the wolf.* Pegasus Books.

Rubin, D. I., & Kazanjian, C. J. (2011). "Just another brick in the wall": Standard-ization and the devaluing of education. *Journal of Curriculum and Instruction, 5*(2), 94–108.

Rukeyser, M. (1968). *The speed of darkness.* Random House.

Rutherford, J. (Ed.). (1990). The third space. Interview with Homi Bhabha. In *Ders. (Hg): Identity: Community, culture, difference* (pp. 207–221). Lawrence and Wishart.

Samuel, R. (2019). *More than a few cells* [Pencil]. Henrietta Lacks Art Show, Evans High School, Evans, GA.

Schubert, W. H. (2009). *Love, justice, and education: John Dewey and the utopians.* Information Age.

Schultz, B. D. (2008). *Spectacular things happen along the way: Lessons from an urban classroom.* Teachers College Press.

Schwab, J. J. (1978). *Science curriculum and liberal education: Selected essays.* University of Chicago Press.

Senior, A. (2011). Haunted by Henrietta: The archive, immortality, and the bio-logical arts. *Contemporary Theater Review, 21*(4), 511–529.

Serres, M. (2007). *The parasite* (L. R. Schehr, Trans.). University of Minnesota Press. (Original work published 1982)

Serres, M. (1997). *The troubadour of knowledge.* Ann Arbor, MI: University of Michi-gan Press. (Original work published 1991)

Serres, M. (2012). *Biogea* (R. Burks, Trans.). Univocal. (Original work published 2010)

Serres, M., & Latour, B. (1996). *Conversations on science, culture and time* (R. Lapi-dus, Trans.). University of Michigan Press. (Original work published 1990)

Shelley, M. (1996). *Frankenstein.* Norton. (Original work published 1818)

Simpson, A. (2021). *The face behind the healing* [Pencil drawing]. Henrietta Lacks Art Show, Evans High School, Evans, GA

Skloot, R. (2010). *The immortal life of Henrietta Lacks.* Broadway Paperbacks.

Skloot, R. (2013, March 23). *New York Times.* http://www.nytimes.com/2013/03/24/opinion/sunday/the-immortal-life-of-henrietta-lacks-the-sequel.html?_r=0

Slattery, P. (2013). *Curriculum development in the postmodern era: Teaching and learn-ing in an age of accountability.* Routledge.

Snow, C. P. (2013). *The two cultures and the scientific revolution.* Cambridge Univer-sity Press. (Original work published in 1959)

Spring, J. (2013). *Common Core: A story of school terrorism.* Create Space Indepen-dent.

Trimble, L. (2021). *Never-ending story* [Multimedia drawing]. Henrietta Lacks Art Show, Evans High School, Evans, GA.

UCLA Institution for Society and Genetics. (2015, June 1). http://socgen.ucla.edu/people/hannah-landecker/

Unit at Tuskegee helps polio fight: Corps of Negro scientists has key role in evalu-ation of Dr. Salk's vaccine. (1955, January 10). *New York Times,* p. 25.

Van Valen, L. (1991). HeLa a new microbial species. *Evolutionary Theory, 10*(2), 71–74.

Virilio, P., & Lotringer, S. (2002). *Crepuscular dawn.* Semiotext(e).

Virillo, P. (1989). *War and cinema: The logistics of perception.* (P. Camiller, Trans). Verso Book. (Original work published in 1984)

Wald, P. (2012a). American studies and the politics of life. *American Quarterly, 64*(2), 185–204.

Wald, P. (2012b). Cells, genes, and stories: HeLa's journey from labs to literature. In K. Wailoo, A. Nelson, & C. Lee (Eds.), *Genetics and the unsettled past: The collision of DNA, race, and history* (pp. 247–265). Rutgers University Press.

Wallace, K. (2015, April 24). *Parents all over U.S. 'opting out' of standardized student testing.* http://www.cnn.com/2015/04/17/living/parents-movement-opt-out-of-testing-feat/

Walls, L. D. (2003). *Emerson's life in science: The culture of truth.* Cornell University Press.

Walls, L. D. (2011). *The passage to the cosmos: Alexander Von Humbolt and the shaping of America.* University of Chicago Press.

Wang, H., & Yu, T. (2006). Beyond promise: Autobiography and multicultural education. *Multicultural Education, 13*(4)29–35.

Weaver, J. A. (2001). Introductions (post)modern science (education): Propositions and alternative paths. In J. A. Weaver, M. Morris, & P. Appelbaum (Eds.), *(Post)modern science (education): Propositions and alternative paths* (pp. 1–22). Peter Lang.

Weaver, J. A. (2010). *Educating the posthuman: Biosciences, fiction, and curriculum studies.* Sense.

Weaver, J. A. (2015). To what future do the posthuman and posthumanism (re)turn us: Meanwhile how do I tame the lingering effects of humanism? In N. Snaza & J. A. Weaver (Eds.), *Posthumanism and educational research* (pp. 182–194). Taylor & Francis.

Weaver, J. A., & Anijar, K. (2001). Part three: Pedagogies of the cultural studies of science. In J. A. Weaver, M. Morris, & P. Appelbaum, *(Post) modern science (education): Propositions and alternative paths* (pp. 243–248). Peter Lang.

Weaver, J. A., Anijar, K., & Daspit, T. (Eds.). (2004). Introduction. In *Science fiction curriculum, cyborg teachers, & youth cultures* (pp. 1–17). Peter Lang.

Weinstein, M., Gleason, S., & Blades, D. W. (2014). *Alternate powers: Deframing the STEM discourse.* Paper presented at the 3rd International Conference of STEM in Education, Vancouver, British Columbia.

Whitman, W. (2011). *Leaves of grass: The original 1855 edition.* Dover Thrift Edition.

Wiesel, E. (1992). *The Nazi doctors and the Nuremberg code: Human rights in human experimentation.* Oxford University Press.

Wilkerson, K. (2015). *HeLa* (K. Wilkerson, Performer). Evans High School, Evans, GA.

Wolfe, G. C. (Director). (2016). *The immortal life of Henrietta Lacks* [Motion Picture].

Wolff, T. (2003). *Old school.* Vintage.

Wulf, A. (2015). *The invention of nature.* Alfred A. Knopf.

Yang, M. (2014, October 4). *Henrietta Lacks* (M. Yang, Performer). Evans High School, Evans, GA.

Zielinski, S. (2011, October 7). *Five historic female mathematicians you should know: Albert Einstein called Emmy Noether a "creative mathematical genius."* http://

www.smithsonianmag.com/science-nature/five-historic-female-mathemati-cians-you-should-know-100731927/?page=5

Zinskie, C. D., & Rea, D. W. (2016). Every student succeeds act (ESSA): What it means for educators and students at risk. *National Youth-At-Risk-Journal, 2*(1), 1–9.

ABOUT THE AUTHOR

Dana Compton McCullough is a high school science teacher at Evans High School in Evans, Georgia. She earned her doctorate in curriculum studies at Georgia Southern University in 2016. Her dissertation, "A Conspiracy to Resurrect Life and Social Justice in Science Curriculum with Henrietta Lacks: A Play," was written to bring awareness to the importance of Henrietta Lacks in cell and molecular biology and the need for science educators to understand stories behind the science and recognize that science instruction must address matters of social justice. She has presented her research at conferences including the American Educational Research Association, and National Science Teacher Association. Her work with students has been featured in *Teaching Tolerance Magazine*, and *Random House Incorporated Web Magazine*.

Printed in the United States
by Baker & Taylor Publisher Services